THE
MILLENNIUM
PROBLEMS

Also by Keith Devlin

THE MILLENNIUM PROBLEMS

The Seven Greatest Unsolved Mathematical Puzzles of Our Time

KEITH DEVLIN

Basic Books
A Member of the Perseus Books Group

Published by Basic Books,
A Member of the Perseus Books Group

Designed by Bookcomp, Inc.

A cataloging-in-publication record for this book is available from the Library of Congress.
ISBN 0-465-01729-0
02 03 04 / 10 9 8 7 6 5 4 3 2 1

CONTENTS

PREFACE

In May 2000, at a highly publicized meeting in Paris, the Clay Mathematics Institute (CMI) announced that seven $1 million prizes were being offered for the solutions to each of seven unsolved problems of mathematics—problems that an international committee of mathematicians had judged to be the seven most difficult and most important in the field today. The announcement caused quite a stir, and for several weeks media interest was high. As a mathematician who writes books and articles for a lay audience and appears regularly on radio, I was approached by numerous journalists and radio producers asking for background or commentary. I was also approached by several editors interested in a book on the topic, among them Bill Frucht at Basic Books. Bill had become a good friend (and something of a hero of mine for his editorial abilities) through our work together on my previous book for a lay audience, *The Math Gene*. So I jumped at the chance to work with him again, and began at once to do the considerable research the book would require.

A short while later, Arthur Jaffe, the president of the Clay Institute, asked me if, together with fellow mathematics popularizer Ian Stewart, I would write general introductions to the seven Millennium Problems for the official book on the problems that the Clay Institute was preparing in conjunction with the American Mathematical Society. After checking that the two books would not conflict unduly, I agreed. The official CMI book consists primarily of detailed and accurate descriptions of the seven problems, each written by one of the world experts on that particular problem. With $1 million at stake for each problem, the CMI book also has the legal responsibility of stating each problem with sufficient precision that judges may determine whether a proposed solution actually meets the problem's criteria. (These problems are not remotely comparable to performing a long division sum or solving a quadratic equation, and in some cases it takes considerable effort simply to understand the individual terms that appear in the statement of the problem.) Ian and I were asked to provide short introductory accounts of the problems to make the book more accessible to mathematicians and more useful to journalists and lay readers who were interested in consulting the "official book" on the problems.

The book you have in your hands is quite different. For the most part, I do not aim at a detailed description of the problems. It is just not possible to describe most of them accurately in lay terms—or even in terms familiar to someone with a university degree in mathematics. (That alone tells you something about the nature of these problems.) Rather, my goal is to provide the background to each problem, to describe how it arose, explain what makes it particularly difficult, and give you some sense of why mathematicians regard it as important.

The official CMI book, then, starts where mine ends. Any reader of this book who wishes to try to solve one of the Clay Problems should, as the very first step, read the definitive description of it in the CMI book. (If you cannot understand that book, you won't be able to solve the problem. The Millennium Prize Competition is like the Super Bowl: It's not a contest for amateurs.) Here I am writing not for those who want to tackle one of the problems, but for readers—mathematician and non-mathematician alike—who are curious about the current state at the frontiers of humankind's oldest body of scientific knowledge. After three thousand years of intellectual development, what are the limits of our mathematical knowledge?

To read my book, all you need by way of background is a good high school knowledge of mathematics. But on its own, that won't be enough. You will also need sufficient interest in the topic. This second prerequisite is more important than the first. I knew from the start that no matter how hard I tried, I could not make this book an easy read. The Millennium Problems are the hardest and most important unsolved mathematics problems in the world; they have resisted numerous attempts at solution, over many years, by the best mathematical minds around. Even achieving a layperson's appreciation of what they are *about* takes considerable effort. I believe the effort, however, is worthwhile. Aren't all pinnacles of human achievement of interest?

By good fortune, you have one additional means of support in trying to come to grips with the Millennium Problems. As part of the efforts the CMI undertook to promote the competition, I worked with Arthur Jaffe and television producer David Stern on a twenty-minute television film that gives the same kind of brief introductory account of the problems that Stewart and I produced for the CMI volume. You can view that film on the web at the CMI website at www.claymath.org (where you will also find technical descriptions of the problems, written by the individual problem experts).

Clearly, my involvement with the CMI book and video helped me in writing this book. I would like to acknowledge Arthur Jaffe and David Ellwood at the CMI for many helpful conversations. My collaboration with Ian Stewart on the introductory passages for the CMI volume also influenced this book. In the end, however, responsibility for the book you have before you is mine, and the blame for any errors or omissions should be laid at my door.

Let me also express my great gratitude to Bill Frucht, who helped conceive the project in the first place and then struggled (and fought) with me in an attempt to make the book as accessible and interesting as possible, despite the inscrutable nature of much of the material we were working with. Thanks too to my agents, Diana Finch in New York and Bill Hamilton in London, for continuing to persuade publishers around the world that there really are people out there who are fascinated by the activities of that (mostly) quiet and unassuming community that I am so pleased to be part of—the seekers of the only 100% certain, eternal truth there is: the mathematicians.

Keith Devlin
Palo Alto, CA
March 2002

THE
MILLENNIUM
PROBLEMS

The Gauntlet Is Thrown

Curiosity is part of human nature. Unfortunately, the established religions no longer provide the answers that are satisfactory, and that translates into a need for certainty and truth. And that is what makes mathematics work, makes people commit their lives to it. It is the desire for truth and the response to the beauty and elegance of mathematics that drives mathematicians.

—Landon Clay, benefactor of the
Clay Millennium Problems

On 24 May, 2000, in a lecture hall at the Collège de France, in Paris, world-renowned mathematicians Sir Michael Atiyah, of Great Britain, and John Tate, of the USA, announced that a prize of $1 million would be awarded to the person or persons who first solved any one of seven of the most difficult open problems of mathematics. These problems, they said, would henceforth be known as the Millennium Problems.

The $7 million prize money—$1 million for each problem, with no time limit on the solution—had been committed by a wealthy American mutual-fund magnate and mathematics afficionado, Landon Clay. One year earlier, Clay had established the Clay Mathematics Institute (CMI), a nonprofit organization based in his home town of Cambridge, Massachusetts, designed to promote and support mathematical research. The CMI had organized the Paris meeting and would administer the Millennium Prize competition.

The seven problems had been selected over several months by a small group of internationally acclaimed mathematicians, chosen by the CMI's scientific advisory board and led by the Clay Institute's founding director, Dr. Arthur Jaffe. Jaffe, a former president of the American Mathematical Society, holds the Landon Clay Chair in Mathematics at Harvard University. The selection committee had agreed that the seven problems chosen were the most significant unsolved problems of contemporary mathematics. Most mathematicians would concur. The problems lie at the center of major areas of mathematics and have all resisted attempts at solution by many of the best mathematicians in the world.

One of the experts who drew up the list is Sir Andrew Wiles, whose solution of Fermat's Last Theorem six years earlier is surely the only reason for that 330-year-old conundrum not being included. The other experts, in addition to Jaffe, were Atiyah and Tate, who gave the Paris lecture, plus Alain Connes, of France, and Edward Witten, of the United States.

Landon Clay, oddly enough, is not a mathematician. As an undergraduate at Harvard he majored in English. Yet he has funded a mathematics chair at his alma mater, then the Clay Mathematics Institute (whose endowment currently stands at $90 million), and now the Millennium Prizes. These initiatives were, he says, partly a response to what he sees as low public funding for an important subject. By offering a major cash prize and inviting the world's press to the meeting at which the competition was to be announced, Clay ensured that the Millennium Problems—and mathematics in general—would get international media attention. But why travel to Paris to do it?

The answer is history. Exactly one hundred years earlier, in 1900, Paris had been host to a similar event. The occasion was the second International Congress of Mathematicians. On August 8, the German mathematician David Hilbert, an international leader in the field, gave an invited address in which he laid out an agenda for mathematics for the twentieth century. In his lecture, Hilbert listed what he judged to be the 23 most significant unsolved problems in mathematics. The "Hilbert Problems," as they became known, were beacons guiding mathematicians forward into the future.

A few of the problems Hilbert stated turned out to be much easier than he had anticipated, and they were soon solved. Several others were too imprecise to admit a definite answer. But most turned out to be mathematical problems of great difficulty. Solving any of those "genuine" Hilbert Problems brought the solver instant fame among the mathematical community, every bit as significant as the award of a Nobel Prize, but with the added advantage that the successful mathematician did not have to wait for years before he (and all the solvers were male) could enjoy the benefits of his success. The accolades came the moment the mathematical community agreed the solution was correct.

By 2000, all but one of the genuine Hilbert Problems had been solved, and the time was ripe for mathematicians once again to take stock. What were the gold-ring problems at the end of the Second Millennium—the unsolved problems that everyone agreed were the Mount Everests of mathematics?

The Paris meeting was partly an attempt to recreate history, but not entirely. As Wiles has pointed out, the CMI's aim in drawing up the Millennium Problems list is not exactly the same as Hilbert's was. "Hilbert was trying to guide mathematics by his problems," Wiles says. "We're trying to record great unsolved problems. There are big problems in mathematics that are important but where it is very hard to isolate one problem that captures the program." The Millennium Problems, in other words, might not give you much of an idea of where mathematics is going. But they provide an excellent snapshot of where the frontier is today.

The Problems

So what are the Millennium Problems? Such is the state of present-day mathematics that none of them can be properly described without considerable background. That's why you are read-ing a book and not an article. But I can at least give you their names now and provide an initial idea of what they are about.

The Riemann Hypothesis. This is the only problem that remains unsolved from Hilbert's list in 1900. Mathematicians the world over agree that this obscure-looking question about the possi-ble solutions to a particular equation is the most significant un-solved problem in mathematics.

The problem was formulated by the German mathematician Bernhard Riemann in 1859 as part of an attempt to answer one of the oldest questions in mathematics: What, if any, is the pat-tern of the prime numbers among all counting numbers? Around 350 B.C., the famous Greek mathematician Euclid had proved that the primes continue forever; that is, there are infinitely many of them. Moreover, by observation, the primes seem to "thin out" and become less common the higher up you go through the counting numbers. But can you say any more than that? The answer—as we shall see in Chapter 1—is yes. A proof of the Rie-mann Hypothesis would add to our understanding of the prime numbers and the way they are distributed. And that would do far more than satisfy the curiosity of mathematicians; besides having implications in mathematics well beyond the patterns of the primes, it would have ramifications in physics and modern communications technology.

Yang–Mills Theory and the Mass Gap Hypothesis. Much of the impetus for new developments in mathematics comes from sci-ence, in particular from physics. For example, it was the needs of physics that led to the invention of calculus by the mathe-maticians Isaac Newton and Gottfried Leibniz in the seventeenth century. Calculus revolutionized science, by providing scientists with a mathematically precise way to describe continuous mo-tion. But although Newton and Leibniz's methods worked, it took some 250 years for the mathematics behind calculus to be properly worked out. A similar situation exists today with some of the theories of physics that have been developed over the past half century or so. The second Millennium Problem challenges mathematicians once again to catch up with the physicists.

The Yang–Mills equations come from quantum physics. They were formulated almost fifty years ago by the physicists Chen-Ning Yang and Robert Mills to describe all of the forces of nature other than gravity. They do an excellent job. The predictions culled from these equations describe particles that have been observed at laboratories around the world. But while the Yang–Mills theory works in practical terms, it has not yet been worked out as a *mathematical* theory. The second Millennium Problem asks for, in part, that missing mathematical development of the theory, starting from axioms. The mathematics would have to meet a number of conditions that have been observed in the laboratory. In particular, it should establish (mathematically) the "Mass Gap Hypothesis," which concerns supposed solutions to the Yang–Mills equations. This hypothesis is accepted by most physicists, and provides an explanation of why electrons have mass. Proof of the Mass Gap Hypothesis is regarded as a good test of a mathematical development of the Yang–Mills theory. It would help the physicists as well. They can't explain why electrons have mass either; they simply observe that they do.

The P Versus NP Problem. This is the only Millennium Problem that is about computers. Many people will find this surprising. "After all," they will say, "isn't most math done on computers these days?" Well, no, actually it isn't. True, most numerical calculations are done on computers, but numerical calculation is only a very small part of mathematics, and not a typical part at that.

Although the electronic computer came out of mathematics—the final pieces of the math were worked out in the 1930s, a few years before the first computers were built—the world of computing has hitherto generated only two mathematical problems that would merit inclusion among the world's most important. Both problems concern computing as a conceptual process rather than any specific computing devices, although this does not prevent them from having important implications for real computing. Hilbert included one of them as number 10 on his 1900 list. That problem—which asks for a proof that certain equations cannot be solved by a computer—was solved in 1970.

The other problem is more recent. This is a question about how efficiently computers can solve problems. Computer scientists divide computational tasks into two main categories: Tasks of type P can be tackled effectively on a computer; tasks of type E could take millions of years to complete. Unfortunately, most of the big computational tasks that arise in industry and commerce fall into a third category, NP, which seems to be intermediate between P and E. But is it? Could NP be just a disguised version of P? Most experts believe that NP and P are not the same (i.e., that computational tasks of type NP are not the same as tasks of type P). But after thirty years of effort, no one has been able to prove whether or not NP and P are the same. A positive solution would have significant implications for industry, for commerce, and for electronic communications, including the World Wide Web.

The Navier–Stokes Equations. The Navier–Stokes equations describe the motion of fluids and gases—such as water around the hull of a boat or air over an aircraft wing. They are of a kind that mathematicians call partial differential equations. University students in science and engineering routinely learn how to solve partial differential equations, and the Navier–Stokes equations look just like the kinds of equations given as exercises in a university calculus textbook. But appearances can be deceptive. To date, no one has a clue how to find a formula that solves these particular equations—or even if such a formula exists.

This failure has not prevented marine engineers from designing efficient boats or aeronautical engineers from building better aircraft. Although there is no general formula that solves the equations (say, in the way that the quadratic formula solves all quadratic equations), the engineers who design high-performance boats and aircraft can use computers to solve particular instances of the equations in an approximate way. Like the

Yang–Mills Problem, the Navier–Stokes Problem is another case where mathematics wants to catch up with what others, in this case engineers, are already doing.

This talk of "catching up" might give the impression that certain problems are important only to the egos of mathematicians, who don't like being left behind. But to think that would be to misunderstand the way scientific knowledge advances. Because of the abstract nature of mathematics, mathematical knowledge about a phenomenon generally represents the deepest and surest understanding of it. And the more deeply we understand something, the better we can take advantage of it. Just as a mathematical proof of the Mass Gap Hypothesis would be a major advance in physics, so too a solution of the Navier–Stokes equations would almost certainly lead to advances in nautical and aeronautical engineering.

The Poincaré Conjecture. This problem, first raised by the French mathematician Henri Poincaré almost a century ago, starts with a seemingly simple question: How can you distinguish an apple from a doughnut? Yes, all right, this doesn't seem like a question that would lead to a $1 million math problem. What makes it hard is that Poincaré wanted a *mathematical* answer that could be used in more general situations. That rules out the more obvious solutions, such as simply taking a bite of each. Here is how Poincaré himself answered the question. If you stretch a rubber band around the surface of an apple, you can shrink it down to a point by moving it slowly, without tearing it and without allowing it to leave the surface. On the other hand, if you imagine that the same rubber band has somehow been stretched in the appropriate direction around a doughnut, then there is no way of shrinking it to a point without breaking either the rubber band or the doughnut. Surprisingly, when you ask whether the same shrinking band idea distinguishes between four-dimensional analogues of apples and doughnuts—which is what Poincaré was really after—no one has been able to provide an answer. The Poincaré conjecture says that the rubber band idea *does* identify four-dimensional apples.

This problem lies at the heart of topology, one of the most fascinating branches of present-day mathematics. Besides its inherent and sometimes quirky fascination—for instance, it tells you the deep and fundamental ways in which a doughnut is the same as a coffee cup—topology has applications in many areas of mathematics, and advances in the subject have implications for the design and manufacture of silicon chips and other electronic devices, in transportation, in understanding the brain, and even in the movie industry.

The Birch and Swinnerton-Dyer Conjecture. With this problem, we're back in the same general area of mathematics as the Riemann Hypothesis. Since the time of the ancient Greeks, mathematicians have wrestled with the problem of describing all solutions in whole numbers x, y, z to algebraic equations like

$$x^2 + y^2 = z^2$$

For this particular equation, Euclid gave the complete solution—that is to say, he found a formula that produces all the solutions. In 1994, Andrew Wiles proved that for any exponent n greater than 2, the equation

$$x^n + y^n = z^n$$

has no nonzero whole-number solutions. (This was the result known as Fermat's Last Theorem.) But for more complicated equations it becomes extremely difficult to discover whether there are any solutions, or what they are. The Birch and Swinnerton-Dyer Conjecture provides information about the possible solutions to some of those difficult cases.

As with the Riemann Hypothesis, to which it is related, a solution to this problem will add to our overall understanding of the prime numbers. Whether it would have comparable implications outside of mathematics is not clear. Proving the Birch and Swinnerton-Dyer Conjecture might turn out to be important only to mathematicians.

On the other hand, it would be foolish to classify this or any mathematical problem as being "of no practical use." Admittedly, the mathematicians who work on the abstract problems of "pure mathematics" are usually motivated more by curiosity than by any practical consequences. But again and again, discoveries in pure mathematics have turned out to have important practical applications.

Moreover, the methods mathematicians develop to solve one problem often turn out to have applications to quite different problems. This was definitely the case with Andrew Wiles's proof of Fermat's Last Theorem. Similarly, a proof of the Birch and Swinnerton-Dyer Conjecture would almost certainly involve new ideas that will later be found to have other uses.

The Hodge Conjecture. This is another "missing piece" question about topology. The general question is about how complicated mathematical objects can be built up from simpler ones. Of all the Millennium Problems, this is perhaps the one the layperson will have the most trouble understanding. Not so much because the underlying intuitions are any more obscure than for the other problems or because it is believed to be harder than any of the other six problems. Rather, the Hodge Conjecture is a highly technical one having to do with the techniques mathematicians use to classify certain kinds of abstract objects. It arises deep within the subject, at a high level of abstraction, and the only way to reach it is by way of those layers of increasing abstraction. This is why I have put this problem last.

The path to the conjecture began in the first half of the twentieth century, when mathematicians discovered powerful ways to investigate the shapes of complicated objects. The basic idea was to ask to what extent you can approximate the shape of a given object by gluing together simple geometric building blocks of increasing dimension. This technique turned out to be so useful that it was generalized in many different ways, eventually leading to powerful tools that enabled mathematicians to catalogue many different kinds of objects. Unfortunately, the generalization obscured the geometric origins of the procedure, and

the mathematicians had to add pieces that did not have any geometric interpretation at all. The Hodge conjecture asserts that
for one important class of objects (called projective algebraic
varieties), the pieces called Hodge cycles are, nevertheless, combinations of geometric pieces (called algebraic cycles).

Those, then, are the Millennium Problems—the most significant and challenging unsolved problems of mathematics at the
turn of the Third Millennium. If you've drawn any conclusion
at all about them from my descriptions, it's probably that they
seem awfully esoteric.

Why Are the Problems So Hard to Understand?

Imagine for a moment that Landon Clay had chosen to establish his prize competition not for mathematics but for some
other science, say physics, or chemistry, or biology. It surely
would not have taken an entire book to explain to an interested lay audience the seven major problems in one of those
disciplines. A three- or four-page expository article in *Scientific
American* would probably suffice. Indeed, when the Nobel Prizes
are awarded each year, newspapers and magazines frequently
manage to convey the gist of the prizewinning research in a few
paragraphs.

In general you can't do that with mathematics. Mathematics
is different. But how?

Part of the answer can be found in an observation first made
(I believe) by the American mathematician Ronald Graham, who
for most of his career was the head of mathematical research at
AT&T Bell Laboratories. According to Graham, a mathematician is the only scientist who can legitimately claim, "I lie down
on the couch, close my eyes, and work."

Mathematics is almost entirely cerebral—the actual *work* is
done not in a laboratory or an office or a factory, but in the
head. Of course, that head is attached to a body, which might
well be in an office—or on a couch—but the mathematics itself
goes on in the brain, *without any direct connection to something in the physical world.* This is not to imply that other sci-

entists don't do mental work. But in physics or chemistry or biology, the object of the scientist's thought is generally some phenomenon in the physical world. Although you and I cannot get inside the scientist's mind and experience her thoughts, we do live in the same world, and that provides the key connection, an initial basis for the scientist to explain her thoughts to us. Even in the case of physicists trying to understand quarks or biologists grappling with DNA, although we have no everyday experience of those objects, even a nonscientifically trained mind has no trouble thinking about them. In a deep sense, the typical artist's rendering of quarks as clusters of colored billiard balls and DNA as a spiral staircase might well be (in fact are) "wrong," but as mental pictures that enable us to visualize the science they work just fine.

Mathematics does not have this helpful link to reality. Even when it is possible to draw a picture, more often than not the illustration is likely to mislead as much as it helps, which leaves the expositor having to make up with words what is lacking or misleading in the picture. But how can the nonmathematical reader understand those words, when they in turn don't link to anything in everyday experience?

Even for the committed spectator of mathematics, this task is getting harder as the subject grows more and more abstract and the objects the mathematician discusses become further and further removed from the everyday world. Indeed, for some contemporary problems, such as the Hodge Conjecture, we may have already reached the point where the outsider simply can't make the connection. It's not that the human mind requires time to come to terms with new levels of abstraction. That's always been the case. Rather, the degree and the pace of abstraction may have finally reached a stage where only the expert can keep up.

Two and a half thousand years ago, a young follower of Pythagoras proved that the square root of 2 is not a rational number, that is, cannot be expressed as a fraction. This meant that what they took to be *the* numbers (the whole numbers and the fractions) were not adequate to measure the length of the hypotenuse of a right triangle with width and height both equal to 1 unit (which Pythagoras's theorem says will have length $\sqrt{2}$). This discovery came as such a shock to the Pythagoreans that their progress in mathematics came to a virtual halt. Eventually, mathematicians found a way out of the dilemma, by changing their conception of what a number is to what we nowadays call the real numbers. To the Greeks, numbers began with counting (the "natural numbers"), and in order to measure lengths you extended them to a richer system (the "rational numbers") by declaring that the result of dividing one natural number by another was itself a number. The discovery that the rational numbers were not in fact adequate for measuring lengths led later mathematicians to abandon this picture, and instead declare that numbers simply *were* the points on a line! This was a major change, and it took two thousand years for all the details to be worked out. Only toward the end of the nineteenth century did mathematicians finally work out a rigorous theory of the real numbers. Even today, despite the simple picture of the real numbers as the points on a line, university students of mathematics always have trouble grasping the formal (and highly abstract) development of the real numbers.

Numbers less than zero presented another struggle. These days we think of negative numbers as simply the points on the number line that lie to the left of 0, but mathematicians resisted their introduction until the middle of the nineteenth century. Similarly, most people have difficulty coming to terms with complex numbers—numbers that involve the square root of negative quantities, such as $i = \sqrt{-1}$ —even though there is a simple intuitive picture of the complex numbers as the points in a two-dimensional plane.

These days, even many nonmathematicians feel comfortable using real numbers, complex numbers, and negative numbers. That is despite the fact that these are highly abstract concepts that bear little relationship with counting, the process with which numbers began some ten thousand years ago, and even though in our everyday lives we never encounter a concrete example of an irrational real number or a number involving the square root of -1.

Similarly, in geometry, the discovery in the eighteenth century that there were other geometries besides the one that Euclid had described in his famous book *Elements* caused both the experts and the nonmathematicians enormous conceptual problems. Only during the nineteenth century did the idea of "non-Euclidean geometries" gain widespread acceptance. That acceptance came even though the world of our immediate everyday experience is entirely Euclidean.

With each new conceptual leap, even mathematicians need time to come to terms with new ideas, to accept them as part of the overall background against which they do their work. Until recently, the pace of progress in mathematics was such that, by and large, the interested observer could catch up with one new advance before the next one came along. But it has been getting steadily harder. To understand what the Riemann Hypothesis says, the first problem on the Millennium list, you need to have understood, and feel comfortable with, not only complex numbers (and their arithmetic) but also advanced calculus, and what it means to add together infinitely many (complex) numbers and to multiply together infinitely many (complex) numbers.

Now, that kind of knowledge is restricted almost entirely to people who have majored in mathematics at university. Only they are in a position to see the Riemann Hypothesis as a simple statement, not significantly different from the way an average person views Pythagoras's theorem. My purpose in this book, then, is not only to explain what the Riemann Hypothesis says but to provide all of the preliminary material as well.

In most cases, that preparatory material can't be explained in terms of everyday phenomena, the way physicists such as Brian Greene[1] can explain the latest, deepest, cutting-edge theory of the universe—Superstring Theory—in terms of the intuitively simple picture of tiny, vibrating loops of energy (the "strings" of the theory). Most mathematical concepts are built up not from everyday phenomena but from earlier mathematical concepts. That means that the only route to getting even a superficial un-

---

1. See his excellent expository book *The Elegant Universe*.

derstanding of those concepts is to follow the entire chain of abstractions that leads to them.

Still, it's important to remember that mathematicians belong to the same species as you. (Trust me on this.) The Millennium Problems are all, by definition, comprehensible to human minds. The concepts involved and the patterns they deal with are not so much inherently difficult as they are very, very unfamiliar— much as the idea of complex numbers or non-Euclidean geometries would have seemed incomprehensibly strange to the ancient Greeks. Today, having grown familiar with these ideas, we can see how they grow naturally out of concepts the Greeks knew as commonplace mathematics. Perhaps the best way to read this book is to think of the seven problems as the commonplace mathematics of the twenty-fifth century.

Who (Just) Wants to Be a Millionaire?

Will the prospect of winning a million-dollar prize have any real effect on whether one of the Millennium Problems is solved? If the question is whether someone will solve the problem in order to win the prize, the answer is no. This is not an amateur competition any more than the Super Bowl is. To solve one of these problems, you would almost certainly need a Ph.D. in mathematics, be good enough to get a tenured position at one of the top universities in the world, and be prepared to devote many years to an in-depth study of the relevant area(s). All at the cost of not doing who knows what else with your time. Anyone who needs the prospect of a million-dollar prize to persuade him or her to do that with their life simply doesn't have the requisite commitment to mathematical research.

On the other hand, the seven prizes may promote progress in a different way. By drawing attention to those particular problems, the prizes could draw young mathematicians into the areas the problems come from. One of those people could then conceivably go on to solve one of the problems. But such a person will without doubt gain far more pleasure from finding the solution than from winning the cash prize. In this respect, mathematicians are no different from Olympic athletes, who invariably value gold medals far more than the lucrative advertising and merchandising contracts they bring.

Ultimately, mathematicians pursue these problems for the same reason the famous British mountaineer George Mallory gave in answer to the newspaper reporter's question, "Why do you want to climb Mount Everest?": "Because it is there." That answer can be taken as either trite or highly profound. If the reporter really could not see why anyone would want to risk his life trying to climb a mountain, then quite likely Mallory gave him the meaningless answer he deserved. On the other hand, it doesn't take a tremendous understanding of human nature to see that Mallory's words reveal a fundamental part of the human spirit: the urge to break new ground, to run faster, jump higher, or climb higher than anyone before—or, since those opportunities present themselves so rarely, at least to go beyond one's own previous bests.

This is why the Millennium Problems are the Mount Everests of mathematics. Ask any of the mathematicians who have spent years of their professional lives trying to solve one of these problems, and the answer you will get will not be too different from, "Because it is there."

Of course, though we all, I suspect, know what it is like to set our sights on a goal, not many of us take it to the same extreme as a world-class mountaineer. But it is surely easy to understand why someone with the necessary ability might do so. I have met several people who have climbed Everest, and knew one who died on its slopes, and what I saw in them was simply a greater love and passion for the sport than I had in my younger days as a weekend rock climber. It was not a masochistic streak or a macabre death wish that led those people to endure enormous hardship and risk their lives trying to scale tall mountains—or the younger me to climb technically difficult cliff faces for that matter. On the contrary, it was a powerful love of life that drove them to do what they did.

Similarly for mathematics. The world's very best mathematicians, the kinds of people who might stand a chance of solving a Millennium Problem, simply bring to the subject a greater degree of commitment and passion than do those of us who gain our intellectual kicks from wrestling with the lesser challenges of mathematics. For any math lover, from the world expert to the amateur who works on mathematical problems on weekends, there doesn't have to be any reason to try to solve a difficult mathematical problem other than, "Because it is there."

Doing research in mathematics is like trying to find your way to the top of a great mountain. You start in the valleys, where the brush and the trees are so dense it's hard to find your way around or even to know which direction to head in. (You may remember that feeling from your high school math class.) But after you have stumbled around for a while, through the trees you catch a glimpse of a tall, snow-covered peak reaching up to the sky. It looks absolutely beautiful. (Unfortunately, most students in the school math class don't even get to this stage. The few who do generally can't resist climbing the mountain; they become mathematicians.) Even when you know where the mountain is, it's still hard battling your way to its base. You keep making wrong turns and having to backtrack, and often you get discouraged by your lack of progress. But if you persist—and are not afraid to ask for directions—then eventually you find yourself looking up toward the summit.

Now you start to climb. As you go higher, the trees and undergrowth get progressively less dense, which tends to make the going easier. (As any professional mathematician knows, advanced mathematics is often much easier than some of the mathematics that is classified as "elementary.") On the other hand, the air gets thinner (the mathematics becomes more abstract), and that tends to make the ascent harder. What's more, the higher you go, the less likely you are to meet guides who can help you find your way. Eventually, you are out on your own. Now a single slip might lead to a major fall. (One small sign error in an equation could destroy months of subsequent research.)

But if you make it to the top, the sense of accomplishment is immense. All the pain of the climb is forgotten in the moment that first rush of success sweeps over you. And the view is breathtaking. From here, at the top of the mountain, you can look down and can see the way you have come, including all your false steps. You can also get a good sense of the terrain below you. As a result, when you are back in the valley, searching for the next peak to climb, things will probably be a bit easier. The next time you will start with the kind of global understanding that comes only from having scaled a large peak and looked down from the summit.

The seven Millennium Problems are the current Mount Everests of mathematics. Quite what you will be able to see from any one of those seven summits is hard to say with any precision. There can be no doubt, however, that if any one of them is solved, we will be able to see so far that the world cannot possibly be left unchanged. That is the true prize. The $1 million price tag each now carries merely acknowledges that status.

ONE

The Music of the Primes

The Riemann Hypothesis

A sk any professional mathematician to name *the* most important unsolved problem of mathematics and the answer is virtually certain to be, "The Riemann Hypothesis." This hundred-and-forty-year-old conundrum asks whether all the (infinitely many) solutions to a certain equation have a particular form. Thus the answer must be either yes or no. Although the equation looks highly specialized, the problem has deep and profound connections to several parts of mathematics. If a *positive* solution is found—and most mathematicians think that this is how it will turn out, probably within the present century—it will have important implications not just for our understanding of the counting numbers but also for much of mathematics, physics, and some key aspects of modern life. That it is the only problem on Hilbert's list of the most important mathematical challenges in 1900 to make it onto the Millennium Problems list a century later only adds to its allure.

Yet for all that the problem lies within the dense undergrowth of modern, abstract mathematics, it originates with a question nearly as old as mathematics itself: the pattern of the primes.

The concept of a prime number—a number that can be evenly divided only by itself and 1—goes back to the mathematicians of ancient Greece, to whom we in the West owe much of our mathematical tradition. In his great thirteen volume book *Elements*, written around 350 B.C., Euclid devoted many pages to the prime numbers. In particular, he proved that every number bigger than 1 (i.e., every positive counting number bigger than 1) is either itself a prime or else can be written as the product of prime numbers in a way that is unique apart from the order in which the primes are written. For example,

$$21 = 3 \times 7,$$
$$260 = 2 \times 2 \times 5 \times 13.$$

The expressions to the right of the equals signs are the "prime decompositions" of the numbers 21 and 260, respectively. Thus, we can express Euclid's result by saying that every counting number bigger than 1 is either prime or else has a unique (up to changing the order) prime decomposition.

This fact, called the fundamental theorem of arithmetic, tells us that the primes are like the chemist's atoms—the fundamental building blocks out of which all numbers are constructed. Just as knowledge of the unique molecular structure of a substance can tell us a lot about its properties, knowing the unique prime decomposition of a number can tell us a lot about its mathematical properties.

So what does the Riemann Hypothesis say? We have a bit of ground to cover before I can answer that question, and the best place to begin is with the man who posed the problem in 1859: Georg Friedrich Bernhard Riemann. Much of our present conception of the nature of mathematics is due to Riemann.

When a Flower Opens It Makes No Noise

Mathematicians learn to accept that their subject is almost certainly more misunderstood than any other—particularly *pure* mathematics, the subject pursued for its own sake, rather than as part of something else such as physics or engineering (when it is sometimes called applied mathematics). The misunderstanding comes on several levels.

For a start, many citizens are unaware that many of the trappings of present-day life depend on mathematics in a fundamental way. When we travel by car, train, or airplane, we enter a world that depends on mathematics. When we pick up a telephone, watch television, or go to a movie; when we listen to music on a CD, log on to the Internet, or cook our meal in a microwave oven, we are using the products of mathematics. When we go into hospital, take out insurance, or check the weather forecast, we are reliant on mathematics. Without advanced mathematics, none of these technologies and conveniences would exist.

Another misunderstanding is that to most people mathematics is just numbers and arithmetic. In fact, numbers and arithmetic are only a very small part of the subject. To those of us in the business, the phrase that best describes the subject is "the science of patterns." Some of the better known of the many specialties of contemporary mathematics are number theory (the study of patterns of numbers), geometry (the study of patterns of shape), trigonometry (which considers measurement of shapes), algebra (which studies patterns of putting things together), calculus (patterns of continuous motion and change), topology (patterns of closeness and relative position), probability theory (patterns of repetition in random events), statistical theory (patterns of real-world data), and logic (patterns of abstract reasoning).

It's not hard to find the reasons for these common misconceptions. Most of the mathematics that underpins present-day science and technology is at most three or four hundred years old, in many cases less than a century old. Yet the typical high school curriculum covers mathematics that is for the most part at least five hundred and in many cases over two thousand years old. It's as if our literature courses gave students Homer and Chaucer but never mentioned Shakespeare, Dickens, or Proust.

Still another common misconception is that mathematics is mainly about performing calculations or manipulating symbolic expressions to solve problems. But this one is different. Whereas a scientist or engineer—indeed anyone who has studied any mathematics at all at the university level—will not harbor the first two misconceptions, possibly only pure mathematicians are likely to be free of this third misconception. The reason is that until 150 years ago, mathematicians themselves viewed the subject the same way. Although they had long ago expanded the realm of objects they studied beyond numbers and algebraic symbols denoting numbers, they still regarded mathematics as primarily about calculation.

In the middle of the nineteenth century, however, a revolution took place. One of its epicenters was the small university town of Göttingen, in Germany, where the local revolutionary leaders were the mathematicians Lejeune Dirichlet, Richard Dedekind, and Bernhard Riemann. In their new conception of the subject, the primary focus was not performing a calculation or computing an answer, but formulating and understanding abstract concepts and relationships—a shift in emphasis from *doing* to *understanding*. Within a generation, this revolution would completely change the way pure mathematicians thought of their subject. Nevertheless, it was an extremely quiet revolution, recognized only when it was all over. It is not even clear that the leaders knew they were spearheading a major change. The oddly quiet emergence of modern mathematics brings to mind the words of the German writer Wilhelm Raabe, who thirty years later wrote, "When a flower opens it makes no noise."[1]

The 1850s revolution did, after a fashion, eventually find its way into school classrooms in the form of the 1960s "New Math" movement. Unfortunately, by the time the message had made its way from the mathematics departments of the leading universities into the schools, it had been badly garbled. To mathematicians before and after 1850, both calculation and un-

---

1. Wilhelm Raabe, *Alte Nester* ("Old Nests"), published in Braunschweig in 1880.

derstanding had always been important. The 1850 revolution merely shifted the *emphasis* as to which of the two the subject was really about and which was the supporting skill. Unfortunately, the message that reached the nation's school teachers in the 60s was often, "Forget calculation skill, just concentrate on concepts." This ludicrous and ultimately disastrous strategy led the satirist Tom Lehrer to quip, in his song *New Math*, "it's the method that's important, never mind if you don't get the right answer." (Lehrer, by the way, is a mathematician, so he knew what the initiators of the change had intended.) After a few sorry years, "New Math" (which was already over a hundred years old) was dropped from the syllabus.

Bernhard Riemann

Bernhard Riemann was an unlikely revolutionary. Born in September 1826 in the small town of Breslenz in what was then the kingdom of Hanover, he was the second of six children. He was a quiet, shy, sickly individual who remained very much a loner throughout his life. One of the few people who seems to have broken through his shell was his colleague Richard Dedekind, who wrote a biography of Riemann ten years after his death. According to Dedekind, on top of his genuinely poor health Riemann was also a hypochondriac.

Bernhard's father, the local Lutheran pastor, very much wanted his son to become a theologian, but Bernhard was soon to display other talents—and a lack of them. He first attended school in Hanover, where he suffered from chronic homesickness, and then in Lüneburg, which was much closer to home. He was hardly a model pupil. He never could take to the drill of learning Latin, and his performance in German composition was poor. Throughout his life he would find writing difficult. Moreover, he would remember only what interested him.

Mathematics interested him, and from the very beginning his mathematical talents were clear to everyone. On the other hand, his obsessive desire for perfection led to many late assignments, and, to his teachers' frustration, he preferred to work things out for himself rather than bother to see what the textbook said. Recognizing his extraordinary abilities, however, his teachers frequently bent the school rules to enable him to progress through the system to graduation.

The principal at Riemann's high school in Lüneburg, a man named Schmalfuss, reported that he encouraged Riemann's interest in mathematics by lending him advanced books. On one occasion, he gave Bernhard his copy of Adrien-Marie Legendre's 900-page text *Number Theory*. Riemann returned it in less than a week, saying, "This is a wonderful book; I know it by heart."

When he became a university student—he studied at Göttingen and Berlin—Bernhard finally abandoned his father's desire that he study theology and switched to mathematics, in which he obtained his doctorate at the University of Berlin in 1851. By then he had abandoned his initial goal of becoming a high school math teacher and embarked on a lifelong career as a university mathematician. He became an associate professor at Göttingen in 1857 and a full professor there in 1859.

His shyness prevented him from getting to know many people beyond his family and immediate colleagues, but that was enough to bring him into contact with a young lady named Elise Koch, a friend of his sisters. The two fell in love, and married in June of 1862. In the fall of that year Riemann contracted pleurisy, which left him with permanent lung damage. After this, the couple spent most of their lives in the warmer climate of Italy, where their only child, a daughter, Ida, was born. Riemann died in July 1866 in Selasca, on the shores of Lake Maggiore.

Despite the presence on the Göttingen faculty of the famous mathematician Karl Friedrich Gauss, the university during Riemann's undergraduate years was not particularly strong in mathematics, and it appears that Gauss himself, then nearing the end of his career, had little influence on him. In 1847, after completing his first year at Göttingen, Riemann transferred to Berlin University, where he studied under a number of world-class mathematicians, notably Jakob Steiner, Carl Jacobi, Gotthold Eisenstein, and Lejeune Dirichlet. The last, in particular, influenced Riemann enormously. The mathematician Felix Klein wrote:

> Riemann was bound to Dirichlet by the strong inner sympathy of a like mode of thought. Dirichlet loved to make things clear to himself in an intuitive substrate; along with this he would give acute, logical analyses of foundational questions and would avoid long computations as much as possible. His manner suited Riemann, who adopted it and worked according to Dirichlet's methods.[2]

Throughout his career Riemann worked primarily in an intuitive way, never exhibiting an appetite for the rigorous logical argument required to make his results watertight. Although this approach infuriated the older generation of mathematicians, for whom calculation was of central importance, it was essential to Riemann's eventual success. While for many mathematicians "intuitive work" can be hit-or-miss, Riemann's mathematical intuitions were incredibly acute, and his results generally turned out to be correct.

In 1849 Riemann returned to Göttingen to work on his doctorate, supervised, at least officially, by Gauss. After Riemann successfully presented his thesis on 16 December 1851, Gauss in his official report described the work as exhibiting "a gloriously fertile originality."

At the time Riemann was completing his thesis, the mathematical revolution was approaching its peak. The conceptual, abstract approach pioneered by Dirichlet was starting to replace the old computational/algorithmic view. "Thinking in concepts" (*Denken in Begriffen*) was what this new generation of mathematicians was about. Mathematical objects were no longer thought of as given primarily by formulas, but rather as carriers of conceptual properties. Proving was no longer a matter of transforming terms in accordance with rules, but a process of logical deduction about concepts.

---

2. Felix Klein, *Development of Mathematics in the Nineteenth Century*, Math-Sci Press, 1979.

Among the new concepts that the revolution embraced are many that are familiar to today's university mathematics student; *function*, for instance. Prior to Dirichlet, mathematicians were used to the fact that a formula such as

$$y = x^2 + 3x - 5$$

specifies a rule that produces a new number y from any given number x. For example, given the number $x = 4$, the formula produces the number $4^2 + (3 \times 4) - 5 = 23$. Dirichlet said forget the formula and concentrate on what the function does. A *function*, according to Dirichlet, is any rule that produces new numbers from old. The rule does not have to be specified by an algebraic formula. In fact, there's no reason to restrict your attention to numbers. A function can be any rule that takes objects of one kind and produces new objects from them. According to this new conception, the rule that associates with each country in the world its capital city is a bona fide function (albeit a nonmathematical one).

Mathematicians began to study the properties of *abstract* functions, specified not by some formula but by their behavior. For example, does the function have the property that when you present it with different starting values it always produces different answers? (This property is called injectivity.) The capital cities function has this property (different countries have different capital cities), but the numerical function $y = x^2$ does not (since, for example, $x = 2$ and $x = -2$ both give the same answer $y = 4$).

This approach was particularly fruitful in the development of calculus, where mathematicians studied the properties of continuity and differentiability of functions as abstract concepts in their own right.[3] In France, Augustin Cauchy developed his famous epsilon–delta definitions of continuity and differentiability—the "epsilontics" that to this day cost each new generation of mathematics students so much effort to master. Cauchy's con-

---

3. See later in the chapter for an explanation of differentiation.

tributions in particular indicated a new willingness of mathematicians to come to grips with the concept of infinity. In a lecture attended by Dedekind, Riemann spoke of their having reached "a turning point in the conception of the infinite."

(For readers with sufficient mathematical background, let me mention also that in 1829 Dirichlet deduced the representability by Fourier series of a class of functions *defined by concepts.* In a similar vein, in the 1850s, Riemann defined a complex function *by its property of differentiability*, rather than a formula, which he regarded as secondary. And Riemann's Göttingen colleague Richard Dedekind examined the new concepts of ring, field, and ideal—each of which was defined as a collection of objects endowed with certain operations. Gauss's residue classes were a forerunner of this approach—now standard—whereby a mathematical structure is defined as a set endowed with certain operations, whose behaviors are specified by axioms.)

Like most revolutions, this one has its origins long before the main protagonists came on the scene. The Greeks had certainly shown an interest in mathematics as a conceptual endeavor, not just calculation, and in the seventeenth century, Gottfried Leibniz (one of the two independent inventors of calculus—the other was Isaac Newton) thought deeply about both approaches. But for the most part, until Riemann and his colleagues came along, mathematics remained primarily a collection of procedures—algorithms—for solving problems.

The problem that today tops the mathematicians' "Most Wanted" list belongs to the branch of mathematics known as analytic number theory. In this remarkable subject, pioneered by Dirichlet as part of the new approach to mathematics, methods of calculus are used to obtain results about the positive whole numbers. Oddly enough, the 1859 paper in which Riemann posed the problem, titled *On the number of primes less than a given magnitude*, was his only publication that dealt with number theory. He wrote it, some say in honor of Gauss (who was an acknowledged master in number theory), to mark his (Riemann's) election to the Berlin Academy earlier that year. It is sketchily written, more like hastily scribbled notes than a published paper. In it he discusses various ideas and methods that might be useful in understanding the distribution of the primes. The problem that would later bear his name seems to have been put in almost as an afterthought—a throwaway remark about the possible solutions to a certain equation that Riemann had been unable to solve. In a letter he subsequently wrote to his colleague Karl Weierstrass, Riemann acknowledged the paper's provisional nature, saying at one point:

> Of course, it would be desirable to have a rigorous proof of this [*the Riemann Hypothesis*]; in the meantime, after a few perfunctory vain attempts, I temporarily put aside looking for one, for it seemed unnecessary for the next objective of my investigation.

Thus did the saga begin.

How Many Primes Are There?

Among the small numbers, primes are very common. Of the numbers 2 to 20, the numbers 2, 3, 5, 7, 11, 13, 17, 19 are prime, a total of eight out of nineteen. The remaining numbers, 4, 6, 8, 9, 10, 12, 14, 15, 16, 18, 20 are all "composite" numbers; that is, they are not prime, since each can be evenly divided by some smaller number (apart from 1).

As you look at larger and larger numbers, however, the primes appear to thin out. While there are 5 primes below 10, there are only 24 below 100 and just 168 below 1000. We can express these figures as the average rate at which primes appear. They crop up at an average rate of 5/10 (or 0.5) below 10, 24/100 (or 0.24) below 100, and 168/1000 (0.168) below 1000. These figures can be thought of as "densities". To calculate the density D_N of the primes less than a number N, you simply take the number $P(N)$ of primes beneath N and divide by N. Thus

$$D_N = \frac{P(N)}{N}$$

Here are the densities of primes for the ranges 1 to 10, 1 to 100, 1 to 1,000, 1 to 10,000, 1 to 100,000, and 1 to 1,000,000:

N :	10	100	1,000	10,000	100,000	1,000,000
D_N :	0.5	0.24	0.168	0.123	0.096	0.078

The farther out we go, the smaller the density becomes. Does this thinning continue, or do we reach a point where it reverses itself and we find lots of primes? Or do we perhaps reach a point where there are no more primes at all? Is there any kind of pattern to the way the primes occur among the totality of all counting numbers? These questions, and others like them, fascinated the ancient Greeks and have tantalized mathematicians ever since.

Euclid himself answered one of them. He proved that the primes continue for ever—there are infinitely many of them. His short and ingenious proof is frequently given in present-day university mathematics classes as an example of abstract mathematical reasoning. (You will find the proof at the end of this chapter, in an appendix.) Other questions about the pattern of the primes have proved more intractable. Two that have attracted a lot of attention, but have so far resisted all attempts at a solution, are the twin primes conjecture and the Goldbach conjecture.

The twin primes conjecture asks if there are infinitely many "twin pairs" of primes —primes that are just 2 apart, such as 11 and 13 or 17 and 19. Computer searches have found many examples of such pairs. The largest discovered to date is the pair

$$4,648,619,711,505 \times 2^{60,000} \pm 1$$

These two primes each have 18,075 digits. They were discovered in the year 2000 using a computer.

Goldbach's conjecture, raised by the amateur mathematician Christian Goldbach in 1742, is that every even number greater than 2 is a sum of two primes. For example,

$$4 = 2 + 2$$
$$6 = 3 + 3$$
$$8 = 3 + 5$$
$$10 = 5 + 5$$
$$12 = 5 + 7$$

Computer searches have verified Goldbach's conjecture for all even numbers as far as 400 trillion (4×10^{14}, completed in 2000), but the conjecture itself remains unproved. The closest anyone has come was in 1966, when the Chinese mathematician Jeng-Run Chen proved that from some number N on, every even number greater than 2 is either a sum of two primes or else a sum of one prime and a product of two primes. (Curiously, Chen's argument does not tell you what N is; it simply shows that such a number exists.)

One of the deepest observations about the pattern of the primes was first made by Riemann's Ph.D. advisor, Karl Friedrich Gauss. Born in 1777, Gauss was a genuine child prodigy. When he was just three years old, he was able to do the payroll calculations for his father's building firm. In elementary school he amazed his teacher by finding the sum of the first 100 numbers in a few minutes. The teacher had set the problem thinking it would keep the class quiet for a considerable time, but Gauss noticed a brilliant shortcut. Suppose you write the sum down twice, he said, once in ascending order, the second time in descending order, directly beneath the first, like this:

1	+	2	+	3	+	4	+	5	+	⋯	+	96	+	97	+	98	+	99	+	100
100	+	99	+	98	+	97	+	96	+	⋯	+	5	+	4	+	3	+	2	+	1

Each vertical pair adds up to 101. Altogether, there are 100 vertical pairs. So the sum of all the numbers in both lines is $100 \times 101 = 10100$. But this will be exactly twice the sum of either one of the rows (since each row contains exactly the same numbers). So the sum of the numbers in either of the two rows will be one-half of 10,100, or 5,050. And that's the answer to the teacher's question.

In 1791, when he was just 14 years old, Gauss noticed that the prime density $D_N = P(N)/N$ is approximately equal to $1/\ln(N)$, where $\ln(N)$ is the natural logarithm of N.[4] As far as Gauss could tell, the bigger N got, the better this approximation became. He conjectured that this was not just an accident, and that by making N sufficiently large, the density D_N could be made as close as you please to $1/\ln(N)$. Gauss was never able to prove his conjecture. This was finally achieved—using some decidedly heavy-duty mathematics—in 1896 by the Frenchman Jacques Hadamard and the Belgian Charles de la Vallée Poussin, working independently. Their result is known today as the Prime Number Theorem.

There are at least two fascinating aspects of this result. First, it demonstrates that, despite the seemingly random way that the primes crop up, there is a systematic pattern to the way they thin out. The pattern is not apparent if you look at an arbitrary finite stretch of the numbers. No matter how far out along the numbers you go, you can find clusters of several primes close together as well as stretches as long as you like in which there are no primes at all. Nevertheless, when you step back and look at the entire sequence of counting numbers, you see that there is a very definite pattern: The larger N becomes, the closer the density D_N gets to $1/\ln(N)$.

The second and far more significant feature of the Prime Number Theorem is the nature of the pattern of the primes that it uncovers. The counting numbers are discrete objects, invented (some prefer to say discovered) by our ancestors some 8,000 years ago as a basis for trading. The natural logarithm function was invented by sophisticated mathematicians a mere two hundred years ago. It is not discrete; rather, its definition depends upon a detailed analysis of infinite processes, and forms part

---

4. The natural logarithm should be familiar to anyone who has taken a first course in calculus. Unfortunately, it would be too great a digression to explain it here, but if you read on you will learn a bit more about it, including its graph, shown in Figure 1.2.

of the discipline sometimes called advanced calculus and sometimes called real analysis. One of several equivalent definitions
of ln(x) is as the inverse to the exponential function e^x.

If you wanted to represent the prime numbers on a graph,
the most obvious way would be to mark a point at each prime
number on the x-axis, as in Figure 1.1.

The graph of the function ln(x), on the other hand, is a
smooth, continuous curve, as shown in Figure 1.2. The question is this: Why is there a connection between the irregularly
spaced points on the x-axis in Figure 1.1 and the smooth curve
shown in Figure 1.2? How is it that the function ln(x) can tell
us something about the pattern of the primes?

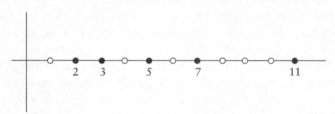

Figure 1.1. The prime numbers displayed graphically.

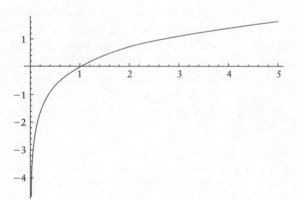

Figure 1.2. The natural logarithm function, ln(x).

The Topography of the Numbers

To most outsiders, doing mathematics involves learning a disjointed collection of rules and techniques for solving various kinds of problems. When faced with a mathematician saying, "Oh, it's obvious, you just do this, then this, and then the answer comes out like this," the average person just assumes that it takes a peculiar kind of brain to do math. That is not the case.[5] What makes it possible for the mathematician to see what to do on such occasions is that she or he sees an underlying structure to the problem domain. When you can see that structure, it often is obvious what to do next.

Imagine, by way of analogy, that you are out for a hike in the forest and you get hopelessly lost. Which way do you proceed? If you look only at what is immediately around you, down on the forest floor, you will have to rely on guesses. A far better approach is to climb a tall tree or make your way to the top of a nearby rise, from where you can see the overall lay of the land. When you see how your immediate area relates to the rest of the terrain, you can better decide which way to go.

Similarly in mathematics. Mathematical knowledge is not a collection of isolated facts. Each branch is a connected whole, and there are links between many of the branches. Think of it as an undulating landscape, much of it heavily forested and shrouded in mist. Trying to find your way around by trial and error is unlikely to get you to your destination. It helps to know as much of the overall topography as possible so you can find the best route to your destination, circumventing impassable rivers and avoiding impossible mountains and dangerous cliffs. How do you achieve that global knowledge? You start out by exploring your immediate vicinity, looking for rises or tall trees

---

5. Much of my book *The Math Gene* is taken up with an argument to show that there is no such thing as a "math brain." Rather, mathematical thinking is in many ways an unnatural activity, and the trick to being able to do mathematics is to approach mathematical problems in such a way that an ordinary human brain can handle them, using mental abilities that were acquired in order to do quite different and for the most part ordinary things.

to climb so you can get a better view. With luck, you will see some mountains. Climb one of them and you can get a good view for many miles in all directions. The higher the mountain, the better the view, of course, but the harder it will be to reach the top. Those individuals who seem to just "know" how to solve math problems have simply spent enough time exploring the mathematical landscape to have developed a good sense of the terrain.

Professional mathematicians employ the landscape analogy in a second, more immediate way as well: They do a great deal of abstract mathematics by viewing it geographically. For example, Figure 1.3 is a computer-generated picture of the landscape corresponding to the mathematical formula $z = \sin(xy)$. The formula is purely abstract and requires training and effort to understand; the geometric figure, on the other hand, tells us a great deal at a glance.

Mathematicians of the post-1850s revolution era began to realize that the best—and maybe the only—way to understand some of the deeper properties of the counting numbers was to view them in the appropriate topographic (i.e., geometric) setting. That setting was what is known as the complex plane.

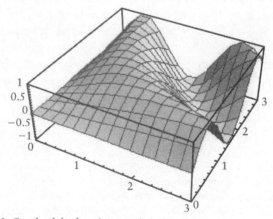

Figure 1.3. Graph of the function $z = \sin(xy)$.

Much of the early development of mathematics can be thought of as driven by the development of number systems applicable to increasingly sophisticated mathematical problems. The most basic numbers of all, the counting numbers, have their origins in what is now the Middle East around 5,000 to 8,000 B.C. The ancient Sumerian civilization introduced them to support commerce. The counting numbers—known to present-day mathematicians as the natural numbers—are adequate for counting collections of objects, but not much more than that.

By the time the ancient Greeks started to develop their mathematics around 700 to 500 B.C., the counting numbers had been extended to allow for fractions—what modern mathematicians call the (positive) rational numbers. With fractions we can count or measure parts of wholes. At first, the Greeks thought that the rational numbers were all you needed to measure lengths with total accuracy, but to their great surprise and consternation, they discovered that this was not the case. Some of the lengths they could create geometrically could not be accurately measured using rational numbers.

For instance, if you construct a right-angled triangle whose base and altitude are both 1 unit, the hypotenuse has a length that is not a rational number. By Pythagoras's theorem, the length of the hypotenuse is $\sqrt{2}$. A young Greek mathematician demonstrated that this number cannot be expressed as the ratio of two whole numbers. According to legend, his colleagues took him out to sea and drowned him to try to keep the awful news hidden. Whether or not that is true, the secret did come out, and the discovery proved devastating for much of ancient Greek mathematics, which never fully recovered from the knowledge.

To be able to measure all geometric lengths, mathematicians had to develop a much richer system of numbers that includes both the natural numbers and the rational numbers as well as a whole lot of others. Those numbers are called the real numbers.

Like the natural and rational numbers, the real numbers have an intuitively acceptable geometric picture. We can view the natural numbers as points on a straight line, first 0, then 1, then 2, then 3, then 4, etc. We can think of the rational numbers as intermediate points on this line, with, for example, $\frac{1}{2}$ halfway between 0 and 1 and $2\frac{3}{4}$ three-quarters of the way between 2 and 3, and so on. If in addition we stipulate that this line is continuous—that is, it has no breaks or gaps—then the real numbers constitute all the points on the line.

This picture of the real numbers as the points on a continuous line is intuitively natural, but it turned out to be a bear to understand mathematically. Although the later ancient Greeks used the real numbers, as did all subsequent mathematicians, a full mathematical understanding of the real numbers did not come until the latter half of the nineteenth century.

It was only in the nineteenth century as well that mathematicians finally accepted negative numbers as genuine numbers. Before this, a simple algebraic equation such as

$$x + 5 = 0$$

was regarded as having no solution. (Today, of course, we would say that the solution is $x = -5$.)

Around the same time as mathematicians were first seriously grappling with the issue of negative numbers, in the sixteenth century, they also struggled with equations such as

$$x^2 + 1 = 0$$

Since the square of any real number—positive or negative—is positive, such an equation cannot have a solution. At least, it cannot have a solution that is a real number. Yet there were physically meaningful situations where such equations came up, and in those cases there should be solutions. For instance, in his book *Ars magna*, an early treatment of algebra, the sixteenth-century Italian Girolamo Cardano, generally referred to as Cardan, presented a method for solving any cubic equation. This method is analogous to the formula

$$x = \frac{-b \pm \sqrt{b^2 - 4ac}}{2a}$$

for the solution of a quadratic equation

$$ax^2 + bx + c = 0$$

But whereas the quadratic formula gives the answer right away, Cardan's method for solving a cubic required several steps, involving formulas that generate intermediate results. For some cubic equations, the intermediate values involved the square roots of negative numbers, even though the final solutions to the cubic were real. Cardan said that although it was "manifestly impossible" to find the square root of a negative number, it was permissible to continue with the calculation if the final result was real. This could happen if any occurrence of a square root of a negative number in an intermediate step was subsequently squared. For example, if $\sqrt{-3}$ arose in an intermediate step, squaring it would produce -3, which is real. Cardan called such intermediate results "sophistic," since they had (he thought) no physical meaning. The term "imaginary" for the square root of a negative quantity seems to have been first used by Euler, who wrote in his book *Algebra* in 1770, "All such expressions as $\sqrt{-1}$, $\sqrt{-2}$, etc. are consequently impossible or imaginary numbers."

But what was the nature of these imaginary intermediate results? To make sense of what was otherwise a perfectly acceptable computation, mathematicians after Cardan invented—perhaps "introduced" is a less loaded term—what are now known as the complex numbers. To get the complex numbers, you begin by postulating a new number i that has the property

$$i^2 = -1$$

The number i (in colloquial terms $\sqrt{-1}$ or "the square root of minus 1") is not a real number; that is to say, it is not a point on the real number line. But you can multiply i by any real number k to form a new number ki. The numbers you get in this way are called the imaginary numbers. (That's the reason for the letter i.) For example, $5i$ is an imaginary number. None of the imaginary numbers is on the real number line (except for $0i$, which is just 0). Geometrically, the imaginary numbers form a second line, perpendicular to the real line, as shown in Figure 1.4.

You can add any real number to any imaginary number to give a new number. For example, you can add the real number $1\frac{2}{3}$ to the imaginary number $5i$ to give the new number $1\frac{2}{3} + 5i$. These combinations of a real number and an imaginary number are called complex numbers. Geometrically, the complex numbers are points in a two-dimensional plane whose x-axis is the real number line and whose y-axis is the imaginary number line. For instance, the complex number $1\frac{2}{3} + 5i$ is the point with x-coordinate $1\frac{2}{3}$ and y-coordinate 5. To get to this particular complex number, you start at the origin of coordinates, go along the x-axis $1\frac{2}{3}$ units, and then go vertically upwards 5 units, as shown in Figure 1.5.

Just as every natural number is automatically a rational number (a whole number can be thought of as a fraction with denominator 1) and every rational number is automatically a real number, so too every real number is automatically a complex number: Just add $0i$ to it.

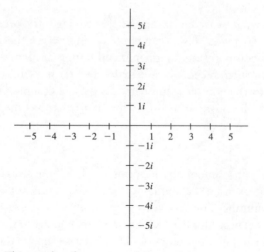

Figure 1.4. The complex plane.

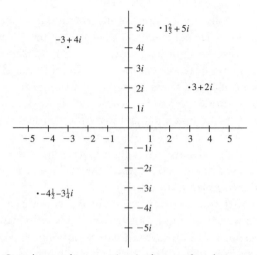

Figure 1.5. Complex numbers as points in the complex plane.

The complex numbers can be added, subtracted, multiplied, and divided, just as the real numbers can. When you add, subtract, multiply, or divide two real numbers, the result is another real number. When you perform these operations on two complex numbers, the result is another complex number. Thus, for doing arithmetic, the complex numbers are very much like the real numbers, except that the rules for complex numbers are a little more complicated.

Despite the growing reliance on complex numbers to solve problems, it was only after Gauss lent his support to their use (the name "complex numbers" is his) that most mathematicians finally accepted them as legitimate mathematical objects. In recent years, the complex numbers have proved invaluable in large parts of mathematics, physics, and engineering. For instance, the standard theory of electric current uses the complex numbers, and as we shall see in Chapter 2, i appears in the most fundamental equation of quantum mechanics. Arguably the most significant arithmetical benefit from working with complex numbers rather than real numbers is that with the complex numbers, every arithmetic (i.e., polynomial) equation has a solution.

Geometrically, the complex numbers are far superior to the real numbers. The real numbers don't really have a geometry; they just sit on a straight line, and all you can do is measure distances along that line. The complex numbers, on the other hand, form a two-dimensional plane, and that means you can do some honest-to-goodness geometry. And how! It turns out that the geometry of the complex plane, most of which was worked out in the nineteenth century, is one of the richest and most beautiful parts of mathematics, having applications way beyond the wildest dreams of the sixteenth-century pioneers who first developed the complex numbers.

Some of the deepest results about the complex plane were discovered by enhancing geometric methods with the powerful techniques of the calculus. Don't panic. Although doing calculus requires considerable advanced training in mathematics, understanding what it is and how it works is fairly straightforward.

A Symphony of Movements

As I have remarked already, mathematics is the science of patterns. The patterns the mathematician studies can be taken from the world around us, or can be extracted from some other discipline or from mathematics itself. A mathematical study of a particular pattern is carried out free of any context, in the most abstract possible way. Different kinds of patterns give rise to different branches of mathematics. Calculus, more precisely differential calculus, is the mathematical study of patterns of continuous motion and change.

Archimedes and Leonardo da Vinci were two of a number of thinkers who tried to describe continuous motion mathematically, but their attempts, while along the right lines, fell short. The key breakthrough came in the middle of the seventeenth century, when Isaac Newton in England and Gottfried Leibniz in Germany independently came up with the right approach. Their idea can be understood in terms of making a movie. As everyone knows, when we watch a movie, the continuous motion we see on the screen is an optical illusion. What we are really seeing is a rapid sequence of still pictures. Each still picture is on the screen for only a twenty-fourth of a second or so, and the difference between two successive pictures is very small. So small, in fact, that when these individual still pictures are projected at a rate of twenty-four frames per second or faster, we cannot see the discrete changes from one picture to the next. We perceive continuous motion.

Calculus works the same way, only mathematically. In calculus, we take some continuous motion and regard it as a sequence of static situations. Each individual frame, as it were, being static, can be analyzed mathematically using the standard techniques of arithmetic and geometry. We analyze the motion by examining the change between pairs of successive static situations, comparing the mathematical analysis of one situation with that of the next. To make the mathematics work, however, we have to imagine that the sequence of static situations advances much faster than the twenty-four frames a second that fools the human eye. We have to imagine that each static situation lasts an infinitesimally short time and that the frames advance at infinite speed.

No movie projector could be built capable of an infinite projection speed, of course. But mathematically, this can be done. In fact, this is exactly what Newton and Leibniz did. They were able to develop mathematical techniques for precisely analyzing the infinitesimally small difference between one situation and the next. Present-day mathematicians call that tiny difference the differential, and the complete theory is called the differential calculus. (See Chapter 4 for more details on differential calculus.)

To make this approach work, it had to be possible to describe the motion being analyzed by means of a reasonably nice mathematical rule or equation. A polynomial equation is fine, as are various other kinds of mathematical rules, including the trigonometric functions $\sin(x)$ (sine of x) and $\cos(x)$ (cosine of x) and the natural logarithm function $\ln(x)$ we met earlier.

A particular example of continuous change is the rise and fall of altitude on a landscape free of any sudden changes of height such as cliff faces or chasms. Given the rule or equation that determines the height at any point (see Figure 1.3 for an example), we can use calculus to calculate how steep the slope is at any point, where the summits are, where the valleys are deepest, whether the slope is getting steeper or leveling off—in short, we can create a comprehensive picture of how the landscape behaves at each point.

In the case of the complex plane, imagine we have a rule f that associates with each complex number $z = x + yi$ a real number $f(z)$. Mathematicians call such a rule f a "real-valued function of a complex variable." We can think of the number $f(z)$ as the "height" of a landscape at the point with coordinates (x, y). Provided the rule has a suitable mathematical description—such as a polynomial equation, a sine, a cosine, or a natural logarithm—the landscape will have no cliffs or chasms, and we can use calculus to study it. (We can regard Figure 1.3 in this way if we view the xy-plane as the complex plane.) If the rule f comes from a problem in some other area of mathematics, then in this way we can represent that problem geometrically and use methods of calculus to study it, and perhaps even solve it.

But this is just the beginning. The real power of using calculus with complex numbers comes when you realize that, since complex numbers are indeed numbers, you can have rules that associate not a real number $r = f(z)$ with a complex number z, but a second complex number $w = f(z)$. Such a rule is called a "complex-valued function of a complex variable," or more simply, a complex function. A complex function associates with each point in the complex plane a sort of generalized "height" that is itself a complex number. The "landscape" you get in this way cannot be visualized, but the mathematics—the algebra, the geometry, and the calculus—still works. Indeed, it works superbly, in many cases giving you a much faster and easier result than you can get with real numbers. It's as if the God of Mathematics has said, "Trust me: In return for having the courage to work with objects that you cannot visualize, I'll reward you by making the mathematics much easier."

This is exactly what happens with the pattern of the primes.

Riemann's Zeta Function

Since the natural numbers are points in the complex plane—all lying on the positive half of the x-axis—when we study properties of the complex plane we can sometimes deduce facts about the natural numbers. Studying the natural numbers by using calculus (and other techniques) to analyze the properties of certain complex functions is a major area of mathematics called analytic number theory. The Riemann Problem is a problem in analytic number theory.

The key to using analytic number theory to study the pattern of the primes is to find a function that provides information about the primes. There are several such. The first was discovered by the famous Swiss mathematician Leonhard Euler, who in 1740 introduced a function that he denoted by the Greek letter zeta (ζ). Euler's "zeta function" associates with any real number s greater than 1 a new real number $\zeta(s)$. To calculate $\zeta(s)$ for a given s, you have to compute the value of an infinite sum:

$$\zeta(s) = \frac{1}{1^s} + \frac{1}{2^s} + \frac{1}{3^s} + \frac{1}{4^s} + \cdots$$

Care should be taken in interpreting this formula, since it uses what I call mathematicians' "dotty notation." Those three dots on the right of the formula may look innocuous, but they are not. They tell us that the sum should be continued forever, following the pattern established by the first four terms. Thus, the fifth term will be $\frac{1}{5^s}$, the sixth $\frac{1}{6^s}$, and so on.

Obviously, you can't work this sum out by adding together infinitely many numbers one at a time. But there are mathematical methods for finding the answer. If s is equal to or less than 1, the sum turns out to be infinite. Perhaps surprisingly, if s is bigger than 1, the sum has a finite answer. (This is why Euler restricted the zeta function to be used only when s is bigger than 1.) Intuitively, what is going on is that if s is bigger than 1, the individual terms $\frac{1}{1^s}, \frac{1}{2^s}, \frac{1}{3^s}$, etc., in the sum grow smaller so fast that, even though you have to add together infinitely many of them, the result is finite. An interesting case is where $s = 2$. Euler himself worked out that $\zeta(2) = \pi^2/6$. (If you want to know more details about how mathematicians compute infinite sums and are curious to see what led Euler to think about the zeta function, see the appendix at the end of this chapter.)

What does the zeta function have to do with the primes? Euler proved that for any (real) number s greater than 1, $\zeta(s)$ is equal to the infinite product

$$\frac{1}{1-(1/2)^s} \times \frac{1}{1-(1/3)^s} \times \frac{1}{1-(1/5)^s} \times \frac{1}{1-(1/7)^s} \times \cdots$$

where the product is taken over all numbers of the form

$$\frac{1}{1-(1/p)^s}$$

for all the prime numbers p.

Despite its rather complicated-looking definition as an infinite sum, the zeta function has some nice mathematical properties. In particular, its graph is smooth (no gaps or sudden jumps), so it can be studied using the methods of calculus.

As a function from real numbers to real numbers, the zeta function is a one-dimensional object, and thus, although it is linked to the primes by Euler's infinite product, it does not have sufficient geometric structure to help you uncover the pattern of the primes. To do that, you need to move up to two dimensions. This is the key step that Riemann made. He replaced the real number s by a complex number z, which made the values $\zeta(z)$ complex numbers as well.

It turns out that there are some complex numbers for which Euler's infinite sum does not give an answer. But a sophisticated mathematical technique known as analytic continuation comes to the rescue. A description of analytic continuation is beyond the scope of this book, but here is the general idea.

Remember that the God of Mathematics is very generous to those having the courage to contemplate complex functions of a complex variable. One of the nice features the G of M provides is this. In cases such as the zeta function, there is an alternative means of calculating the values, *which works for almost all complex numbers*, including all or many of those for which the original formula does not give an answer. For the zeta function itself, the alternative method allows us to calculate $\zeta(z)$ for any complex number z, with the single exception of the number $z = 1$. This process of going from the original definition of the function to the alternative one is called analytic continuation. Since Riemann was the first person who made this switch, the *complex* function ζ is usually referred to as Riemann's zeta function. In his remarkable paper of 1859, Riemann used the zeta function to investigate the pattern of the primes.

His original goal was to prove Gauss's conjecture that for large numbers n, the density of the primes below n (D_n) is approximated by $1/\ln(n)$, the result now known as the Prime Number Theorem. Although he did not achieve that goal, his work did provide a firm link between the prime numbers and the geometry of the complex plane. Moreover, his methods provided the basis of the proof of the Prime Number Theorem that Hadamard and de la Vallée Poussin eventually found in 1896.

The key link between the zeta function and the prime numbers that Riemann found was a close connection between the density function D_n and the solutions to the equation $\zeta(z) = 0$.

Any complex number that solves this equation is said to be a " zero" of the zeta function. In his paper—which was a mere eight pages long—Riemann made a bold conjecture about the zeros of his zeta function. He began by observing that each of the numbers $-2, -4, -6, \ldots$ is a zero. That is to say, $\zeta(z) = 0$ whenever z is a negative even whole number. He then showed that the zeta function must have infinitely many other, complex, zeros besides these real ones. His conjecture was that all of those other zeros have the form $z = \frac{1}{2} + bi$ for some real number b. That is, they all start with $\frac{1}{2}$. In geometric terms, all nonreal zeros of the zeta function lie on the straight line in the complex plane that runs vertically through the point $\frac{1}{2}$ on the x-axis—what is generally referred to as the critical line. See Figure 1.6.

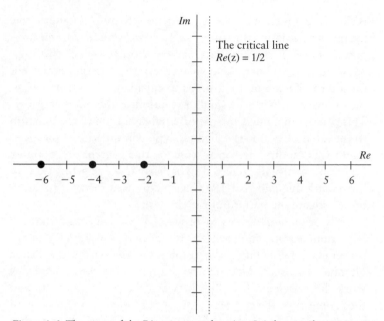

Figure 1.6. The zeros of the Riemann zeta function. It is known that $\zeta(n) = 0$ for all negative even integers n. The Riemann Hypothesis says that the infinitely many other complex numbers z for which $\zeta(z) = 0$ all lie on the critical line $Re(z) = \frac{1}{2}$.

Hadamard and de la Vallée Poussin did not need this conjecture about the zeros—nowadays known as the Riemann Hypothesis—to prove the Prime Number Theorem. The link between the prime numbers and the geometry of the complex plane provided by the Riemann zeta function was enough for that. But Riemann's conjecture, if it is true, has major implications for our knowledge of the prime numbers. Riemann showed that if all the complex (nonreal) zeros of the zeta function have real part equal to $\frac{1}{2}$, then the degree to which the density function D_n differs from the curve $1/\ln(n)$ varies in a systematically random fashion, much as the proportion of heads you get when you toss a coin repeatedly varies from $\frac{1}{2}$. This means that although you can't predict with any accuracy where the next prime will occur, the overall pattern of primes is extremely regular.

The additional information about the pattern of primes that a proof of the Riemann hypothesis would yield would be important not only in mathematics. It could also have major consequences for what has come to be a crucial component of modern life: Internet security.

The Riemann Hypothesis and the World Wide Web

Every time you use an ATM machine at your bank or carry out a business transaction on the Internet, you are depending on the mathematical theory of prime numbers to keep your transaction secure. Here's how.

From the moment people started to send messages to one another, the following issue arose: How can you prevent an unauthorized person who gets hold of the message from understanding what it says? The answer is you encode the message (the technical term is "encrypt") so that only the intended receiver can access the original contents. Julius Caesar used a very simple system to encrypt the messages he sent to his generals commanding the Roman legions across Europe. He simply replaced each letter of the alphabet in each word by another, according to a fixed scheme, such as "replace each letter by the next but one in the alphabet" (with A replacing Y and B replacing Z). A message encrypted this way might look completely unreadable, but to today's cryptanalyst, Caesar's system is easily broken.

These days, with the computer power available to any would-be codebreaker, it is extremely difficult to design a secure encryption system. (Much of the impetus to develop computer technology during the Second World War came from the desire of each side to break the enemy's codes.) If there is any kind of "recognizable" pattern to the encrypted text, a sophisticated statistical analysis using a powerful computer can usually crack the code. So, your encryption system needs to be sufficiently robust to resist computer attack.

These days, encryption systems invariably consist of two components: an encryption procedure and a "key." The former is typically a computer program or, in the most widely used systems, a specially designed computer chip; the key is usually a secretly chosen number. To encrypt a message, the system requires not only the message but also the chosen key. The encryption program codes the message so that the encrypted text can be decoded only with the key. Since the security depends on the key, the same encryption program may be used by many people for a long period of time, and this means that a great deal of time and effort can be put into its design. An obvious analogy is that manufacturers of safes and locks design one type of lock which may be sold to millions of users, who rely upon the uniqueness of their own key—be it a physical key or a secret number combination—to provide security. Just as an enemy may know how your lock is designed and yet be unable to break into your safe, so too the enemy may know what encryption system you are using without being able to break your coded messages.

In early key systems, the message sender and receiver agreed beforehand on some secret key, which they then used to send each other messages. As long as they kept this key secret, the system would (it was hoped) remain secure. An obvious drawback with such an approach was that the sender and receiver had to agree in advance on the key they would use, and since they would clearly not want to transmit that key over any interceptable communication channel, they would have to meet ahead of time and choose the key (or perhaps employ a trusted courier to communicate it). Such a system is obviously unsuitable in many situations. In particular, it won't work in, say, international banking or commerce, where it is often necessary to send secure messages across the world to someone the sender has never met.

In 1975, two mathematicians, Whitfield Diffie and Martin Hellman, proposed a quite new type of encryption system: public key cryptography, in which the encryption method requires not one but two keys—one for encryption and the other for decryption. Such a system is used like this. A new user, say Maria, obtains the standard program (or special computer chip) used by all members of the communication network concerned. She then generates two keys. One of these, her deciphering key, she keeps secret. The other key, the one used for encrypting messages sent to her by anyone else in the network, she publishes in a directory of the network users. To send a message to a network user, all that has to be done is to look up that user's public encryption key, encrypt the message using that key, and send it. To decode the message it is of no help knowing (as anyone can) the encryption key. You need the decryption key, which only the intended receiver knows.

Several specific methods were developed to implement Diffie and Hellman's general scheme. The one that gained most support, and which remains to this day the industry standard, was designed by Ronald Rivest, Adi Shamir, and Leonard Adelman, of the Massachusetts Institute of Technology. It is known by their initials as the RSA system, and is marketed by a commercial data security company, RSA Data Security, Inc., based in Redwood City, California. The secret decryption key used in the RSA method consists (essentially) of two large prime numbers (each having, say, 100 digits) chosen by the user. (The choice of the two primes is made using a computer, not chosen from any published list of primes, to which an enemy might have access. Modern computers can find large primes with ease.) The public encryption key is the product of these two primes. The system's security depends upon the fact that there is no known quick method of factoring large numbers. This means that it is practically impossible to recover the decryption key (the two primes) from the public encryption key (their product). Message encryption corresponds to multiplication of two large primes (an easy task), decryption corresponds to the opposite process of factoring (which is hard). (This is not exactly how the system works. Some moderately sophisticated mathematics is involved.)

At present, the largest numbers that can be factored on a powerful computer in less than a few days have around 90 to 100 digits, so using a key obtained by multiplying together two

100-digit primes, i.e., a number with 200 digits, should make the RSA system extremely secure. But there is a danger. The methods mathematicians use to factor large numbers are not simply trial-and-error searches such as you might use if I asked you to find the prime factors of 221. Doing it that way is fine for fairly small numbers, but could take a powerful computer over a year to factor a 60-digit number. Instead, mathematicians employ some highly sophisticated techniques to find prime factors. The methods they have developed are clever and powerful, and getting steadily more so. Those methods make use of much that we know about prime numbers, and every time there is an advance in our knowledge of the primes, there is always a possibility that it will lead to a new method for factoring numbers.

Since the Riemann Hypothesis tells us so much about the primes, a proof of that conjecture might well lead to a major breakthrough in factoring techniques. Not because we will then know the hypothesis is true. Suspecting that it is true, mathematicians have been investigating its consequences for years. Indeed, some factoring methods work on the *assumption* that it's true. Rather, the fear among the encryption community is that the *methods* used to prove the hypothesis will involve new insights into the pattern of the primes that will lead to better factoring methods.

Clearly, then, with Internet security and large parts of contemporary mathematics hanging in the balance, far more is at stake in the Riemann Problem than the $1 million Millennium Prize.

Is the Riemann Hypothesis True?

Even if Riemann was unable to prove his conjecture, what led him to suspect it in the first place? We will probably never know. The eminent nineteenth-century mathematician Felix Klein said of Riemann's paper, "Riemann must have very often relied on his intuition."

When scholars looked through Riemann's papers after his death, they found the following tantalizing note: "These properties of $\zeta(s)$ (the function in question) are deduced from an expression of it which, however, I did not succeed in simplifying enough to publish it."[6] No one knows what expression Riemann was referring to. It may be related to a result he did establish in his paper: that there is another function $\gamma(s)$ such that for all complex numbers $s \neq 1$,

$$\pi^{-s/2}\gamma(s)\zeta(s) = \pi^{-(1-s)/2}\gamma(1-s)\zeta(1-s)$$

This equation tells us that the values of the zeta function at s are closely related to its values at $1-s$. Putting this another way, the zeta function has a certain symmetry about the critical line (i.e., the vertical line through the point $\frac{1}{2}$ on the real axis). Maybe Riemann intuited that this symmetry forces all zeros to be on the line of symmetry? We don't know.

It was long hoped that Riemann had left a clue buried elsewhere in his notes. In 1932, the mathematician Carl Ludwig Siegel carried out a detailed study of all of Riemann's papers and reported, "No part of Riemann's writings related to the zeta function is ready for publication; occasionally one finds disconnected formulas on the same page; frequently just one side of an equation has been written down; remainder estimates and investigations of convergence are invariably missing, even at essential points."[7] For a while, there was a widespread belief, propagated by Klein and Edmund Landau among others, that Riemann found his results by means of "great general ideas," without using the formal tools of analysis. But Siegel dismisses this.

The fact is, we may never know for sure how Riemann came to his belief. But will we ever know whether it is true?

Using computers, mathematicians have managed to show

---

6. Quoted in Jacques Hadamard, *The Mathematician's Mind*, Princeton University Press (1945, 1973), p. 118. On seeing this quotation, readers familiar with the origin of Fermat's last theorem will doubtless have a feeling of—in the immortal words of Yogi Berra—*déjà vu* all over again.

7. Carl Siegel, Über Riemanns Nachlass zur analytischen Zahlentheorie, *Quellen und Studien zur Geschichte der Mathematik, Astronomie und Physik*, p. 46.

that Riemann's hypothesis is true for the first 1.5 billion zeros of the zeta function. ("First" in the sense of being closest to the real axis.) In most walks of life, supporting evidence on that scale would be regarded as conclusive, but not so in mathematics. All the computer work tells us is that if there is a zero that is not on the vertical line through the real number $\frac{1}{2}$, it will have to be pretty big. But so what? Since the numbers go on for ever, there are plenty of possibilities for there to be a zero that does not satisfy Riemann's criterion. Perhaps there is such a zero, but it is far too big for any computer to handle—ever.

Still, most mathematicians believe Riemann's conjecture is true.

Perhaps the most intriguing result about the Riemann Hypothesis is its connection with quantum physics. In 1972, the American mathematician Hugh Montgomery found a formula that describes the spacing between the zeros of the zeta function along the critical line. Physicists immediately recognized Montgomery's formula as one that would give the spacings between the energy levels of what theoretical physicists call a quantum chaotic system. This raised two possibilities: that a proof of the Riemann Hypothesis might have ramifications in quantum physics, or, conversely, that ideas of quantum physics might lead to a proof of the Riemann Hypothesis. The French mathematician Alain Connes, pursuing the latter possibility, has written down a system of equations that specifies a hypothetical quantum chaotic system that has all the prime numbers built in, and has proved that this system has energy levels corresponding to all the zeros of the zeta function that lie on the critical line. If—and it seems to be a big "if"—he can prove that there are no zeros other than those corresponding to energy levels, he will have proved the Riemann Hypothesis.

If Connes's approach turns out to work, it will be a remarkable illustration of the connection between mathematics and quantum physics, and the first time methods of quantum physics have been used to solve a problem of pure mathematics.

APPENDIX 1

EUCLID'S PROOF THAT THERE
ARE INFINITELY MANY PRIMES

Clearly, Euclid could not show that there are infinitely many primes by listing all of them. He had to work indirectly. What he did was to show that there is no largest prime. Here is a modern version of his proof.

Suppose there were, in fact, a largest prime number. Call it P.

Now, says Euclid, multiply together all the prime numbers from 2 to P, inclusive. No, you don't have to actually perform the computation. How could you? You don't have an actual value for P. Rather, you simply let N denote the result of that computation, whatever it is. That is,

$$N = 2 \times 3 \times 5 \times 7 \times 11 \times \cdots \times P$$

Now look at the number $N + 1$. It is obviously larger than P. Euclid claimed that this number $N + 1$ is a prime number. If he is right, then this will show that there is no largest prime number. Why? Well, look back at what we just said. We started out by supposing (against our better judgment, perhaps) that there was in fact a largest prime number. We decided to call it P. Now, according to Euclid, we have found an even bigger prime number, $N + 1$. Even larger than the largest one? Come off it! This is a logically inconsistent situation. Since we arrived at this state of affairs by supposing that there was a largest prime number, that must be the source of the inconsistency. (The only other things we did were to give that largest prime a name, P, and then specify how the number N was obtained by a straightforward multiplication, and neither of those steps could give rise to an inconsistency.) The conclusion is, therefore, that there is in fact no largest prime number.

So how did Euclid show that the number $N + 1$ is a prime number? This is the step we left out above. He begins by asking what happens if $N + 1$ were not a prime number. In that case, by the fundamental theorem of arithmetic, $N + 1$ must be a product of prime numbers. In particular, $N + 1$ can be divided by some smaller prime number, say M. But M divides evenly into N (because N is the product of all the prime numbers). Hence, when you try to divide M into $N + 1$ you are left with a remainder of 1. Once again we have an inconsistent state of affairs: M divides evenly into $N + 1$ and M leaves a remainder of 1 when you divide it into $N + 1$. And once again, this means that our initial supposition must have been false—in this case the supposition that $N + 1$ was not prime. So (based on the assumption that P is the largest prime) $N + 1$ is indeed a prime number.[8]

---

8. It is not the case that numbers of the form $N + 1$ are always prime. The proof given is based on the (false) assumption that there is no largest prime.

APPENDIX 2

HOW DO MATHEMATICIANS
WORK OUT INFINITE SUMS?

Just how do mathematicians compute infinite sums like the one in the zeta function? Or an even more basic question: How can an infinite sum possibly have a finite answer?

Suppose we start to add together the terms in the infinite sum you get by adding together the squares of all the positive whole numbers:

$$1 + 4 + 9 + 16 + 25 + \cdots$$

Then the successive answers we get are $1 + 4 = 5$, $5 + 9 = 14$, $14 + 16 = 30$, $30 + 25 = 55$, etc. Since these partial sums are growing rapidly, this infinite sum will have an infinite answer. Now take a look at the sum of the reciprocals of the squares of all positive whole numbers:

$$\frac{1}{1} + \frac{1}{4} + \frac{1}{9} + \frac{1}{16} + \frac{1}{25} + \cdots$$

That is,

$$1 + 0.25 + 0.11111 + 0.0625 + 0.04 + \cdots$$

For this infinite sum, the successive partial sums are 1, 1.25, 1.36111, 1.42361, 1.44361, etc. These partial sums are increasing, but by smaller and smaller amounts, and it seems at least a possibility that the infinite sum has a finite answer, possibly between 1 and 2. And in fact this is exactly what turns out to be the case. The sum works out to be about 1.64493. (The exact answer, obtained by a roundabout method, is $\pi^2/6$.)

How do mathematicians go about working out an infinite sum

$$a_1 + a_2 + a_3 + \cdots ?$$

The most basic approach is to look at the successive partial sums

$$a_1 + a_2 + \cdots + a_N$$

for increasing values of N ($N = 1, 2, 3, 4, 5$, etc.), just as we did a moment ago, but instead of simply working out these partial sums, you try to find a pattern in them. For example, in the case of the sum of the reciprocals of the squares of the whole numbers, you can show that as you add more and more terms, the partial sum keeps increasing beyond the value 1.44361 we obtained above, but once the partial sum gets beyond 1.6, the addition of further terms does not alter the first decimal place. The second decimal place continues to change until the partial sum exceeds 1.64, from which point on the addition of still more terms affects only the values of the third and subsequent decimal places. Then the third decimal place settles down to be 4, then the fourth place to be 9, etc. The crucial factor here is that the individual terms are getting small so fast that, even though you have to add infinitely many of them together, one by one the decimal places in the partial sums settle down.

Sometimes, when an infinite series has a particularly simple pattern, you can find a formula for the sum. For example, the infinite sum

$$1 + \frac{1}{2} + \frac{1}{4} + \frac{1}{8} + \frac{1}{16} + \cdots + \frac{1}{2^n} + \cdots$$

has the pattern

$$1 + x + x^2 + x^3 + \cdots + x^n + \cdots$$

(Just take $x = \frac{1}{2}$.) A sum of this form is called a geometric series. If $0 < x < 1$, the geometric series has a finite total. Call it s. Thus

$$s = 1 + x + x^2 + x^3 + x^4 + \cdots$$

If we multiply the entire series through (term-by-term) by x, we get

$$xs = x + x^2 + x^3 + x^4 + x^5 + \cdots$$

This is the original series with the first term missing. So, if we subtract the second series from the first, all terms but the initial 1 in the first series cancel out, leaving

$$s - xs = 1$$

This can be solved for s algebraically to give

$$s = \frac{1}{1 - x}$$

For the particular case where $x = \frac{1}{2}$, this gives the answer $s = 2$.

In general, then, an infinite sum will have a finite answer, provided the individual terms grow smaller at a sufficiently rapid rate. The key question now is, how rapid is "rapid"? For example, if $0 < x < 1$, the terms in a sum of the form

$$s = 1 + x + x^2 + x^3 + x^4 + \cdots$$

definitely grow small sufficiently rapidly to give the finite total we just calculated. How about the sum

$$1 + \frac{1}{2} + \frac{1}{3} + \frac{1}{4} + \frac{1}{5} + \cdots \quad ?$$

Do its terms grow small sufficiently fast for the series to have a finite sum? This particular series is connected with musical harmonics, and as a result mathematicians call it the "harmonic series." It turns out that the harmonic series is a marginal case. Although the individual terms grow smaller, they don't do it fast enough, and as a result the sum is infinite. However, for any exponent s bigger than 1, no matter how small the amount by which s exceeds 1, the sum

$$1 + \frac{1}{2^s} + \frac{1}{3^s} + \frac{1}{4^s} + \frac{1}{5^s} + \cdots$$

is finite—the answer that Euler took for $\zeta(s)$, the value of the zeta function at s.

APPENDIX 3

HOW EULER DISCOVERED
THE ZETA FUNCTION

Now that we know about infinite sums, we can ask how Euler discovered that one particular infinite sum, the zeta function, provides information about the pattern of the primes. Besides being a fascinating question in its own right, the answer will help you to follow the account of the Birch and Swinnerton-Dyer Conjecture in Chapter 6.

Knowing that the harmonic series has an infinite sum, Euler wondered about the "prime harmonic series"

$$PH = 1 + \frac{1}{2} + \frac{1}{3} + \frac{1}{5} + \frac{1}{7} + \frac{1}{11} + \cdots$$

which you get by adding the reciprocals of all the primes. Is its sum finite or infinite?

He began by regarding PH as a subseries of the harmonic series

$$1 + \frac{1}{2} + \frac{1}{3} + \frac{1}{4} + \frac{1}{5} + \frac{1}{6} + \cdots$$

This series has an infinite sum, so it did not allow Euler to do what he wanted to do next. He therefore looked instead at the related sum

$$\zeta(s) = 1 + \frac{1}{2^s} + \frac{1}{3^s} + \frac{1}{4^s} + \frac{1}{5^s} + \cdots$$

you get by raising each term in the harmonic series to the power s. Provided s is bigger than 1, this sum is finite, and so you can split it up into two parts, the first part being all the prime terms, the second all the nonprime terms, like this:

$$\zeta(s) = \left[1 + \frac{1}{2^s} + \frac{1}{3^s} + \frac{1}{5^s} + \cdots \right] + \left[\frac{1}{4^s} + \frac{1}{6^s} + \frac{1}{8^s} + \frac{1}{9^s} + \cdots \right]$$

The idea then is to show that if you were to take s closer and closer to 1, the first sum

$$1 + \frac{1}{2^s} + \frac{1}{3^s} + \frac{1}{5^s} + \cdots$$

increases without bound, and hence that taking $s = 1$,

$$1 + \frac{1}{2} + \frac{1}{3} + \frac{1}{5} + \cdots$$

is infinite.

A key step in this argument was to establish the celebrated equation

$$\zeta(s) = \frac{1}{1 - (1/2^s)} \times \frac{1}{1 - (1/3^s)} \times \frac{1}{1 - (1/5^s)} \times \frac{1}{1 - (1/7^s)} \times \cdots$$

where the product on the right is taken over all terms $\dfrac{1}{1 - (1/p^s)}$ where p is a prime. Euler's idea was to start with the formula for the geometric series we met a moment ago:

$$\frac{1}{1 - x} = 1 + x + x^2 + x^3 + \cdots \qquad (0 < x < 1)$$

For any prime number p and any $s > 1$, we can set $x = 1/p^s$ to give

$$\frac{1}{1 - (1/p^s)} = 1 + \frac{1}{p^s} + \frac{1}{p^{2s}} + \frac{1}{p^{3s}} + \cdots$$

The expression on the left is a typical term in Euler's infinite product, of course, so the above equation provides an infinite sum expression for each term in the infinite product. What Euler did next was multiply together all of these infinite sums to give an alternative expression for his infinite product. Using the ordinary algebraic rules for multiplying (a finite number of finite) sums, but applying them this time to an infinite number of infinite sums, you see that when you write out the right-hand side of the product as a single infinite sum, its terms are all the expressions of the form

$$\frac{1}{p_1{}^{k_1 s} \cdots p_n{}^{k_n s}}$$

where $p_1, \ldots, p_n$ are different primes and $k_1, \ldots, k_n$ are positive integers, and each such combination occurs exactly once. But by the fundamental theorem of arithmetic, every positive integer can be expressed in the form $p_1{}^{k_1 s} \cdots p_n{}^{k_n s}$. Hence the right-hand side of the product is just a rearrangement of the sum

$$1 + \frac{1}{2^s} + \frac{1}{3^s} + \frac{1}{4^s} + \frac{1}{5^s} + \cdots$$

i.e., $\zeta(s)$. (You have to be a bit careful how you do this, to avoid getting into difficulties with infinities. The details are not particularly difficult, but it would take us too far from our path to give the complete argument.)

Now, from our point of view—and indeed from the point of view of the subsequent development of mathematics—it was not so much the fact that the prime harmonic series has an infinite sum that is important, even though it did provide a completely new proof of Euclid's result that there are infinitely many primes. Rather, Euler's infinite product formula for $\zeta(s)$ marked the beginning of analytic number theory.

In 1837, the German mathematician Peter Gustav Lejeune Dirichlet generalized Euler's method to prove that in any arithmetic progression $a, a+k, a+2k, a+3k, \ldots$, where a and k have no common factor, there are infinitely many primes. (Euclid's theorem can be regarded as the special case of this for the arithmetic progression $1, 3, 5, 7, \ldots$ of all odd numbers.) The principal modification to Euler's method that Dirichlet made was to modify the zeta function so that the primes were separated into separate categories, depending on the remainder they left when divided by k. His modified zeta function had the form

$$L(s, \chi) = \frac{\chi(1)}{1^s} + \frac{\chi(2)}{2^s} + \frac{\chi(3)}{3^s} + \frac{\chi(4)}{4^s} + \cdots$$

where $\chi(n)$ is a special kind of function—which Dirichlet called a "character"—that splits the primes up in the required way. In particular, it must be the case that $\chi(mn) = \chi(m)\chi(n)$ for any m, n. (The other conditions are that $\chi(n)$ depends only on the remainder you get when you divide n by k, and that $\chi(n) = 0$ if n and k have a common factor.)

Any function of the form $L(s, \chi)$ where s is a real number greater than 1 and χ is a character is known as a Dirichlet L-series. The Riemann zeta function is the special case that arises when you take $\chi(n) = 1$ for all n.

Mathematicians subsequent to Dirichlet used L-series (in the more general case where the variable s and the numbers $\chi(n)$ are allowed to be complex numbers) to prove a great many results about prime numbers, thereby demonstrating that Dirichlet's series provides an extremely powerful tool for the study of the primes.

A key result about L-series is that, as with the zeta function, they can be expressed as an infinite product over the prime numbers (sometimes known as an Euler product), namely,

$$L(x, \chi) = \frac{1}{1 - (\chi(2)/2^s)} \times \frac{1}{1 - (\chi(3)/3^s)}$$
$$\times \frac{1}{1 - (\chi(5)/5^s)} \times \frac{1}{1 - (\chi(7)/7^s)} \times \cdots$$

(provided the real part of s is not negative), where the product is taken over all expressions of the form

$$\frac{1}{1 - (\chi(p)/p^s)}$$

where p is a prime number.

TWO

The Fields We Are Made Of

*Yang–Mills Theory and
the Mass Gap Hypothesis*

During the twentieth century, physicists made dramatic progress toward answering one of the oldest, most fundamental and tantalizing questions of all: What is the stuff that we, and everything else in the universe, are made of? In doing so, however, they had to make a number of assumptions about the solutions to certain mathematical problems. Since those assumptions, and their consequences, were supported by strong experimental and observational evidence, the scientists are confident about the overall validity of the theory. For the mathematician, however, the physicists' rapid progress has thrown up a major challenge: Work out the math behind the physics. A solution to the second Millennium Problem would be a significant step toward meeting that challenge and would increase our understanding of the nature of matter. This makes the problem the latest step in humanity's long quest to comprehend the universe—a quest that for the last two thousand years has relied heavily on mathematics.

God the Geometer

The ancients believed that the world was made up of four basic elements: earth, water, air, and fire. Foreshadowing the atomic theory of matter of twentieth-century science, around 350 B.C. the ancient Greek philosopher Plato, in his book *Timaeus*, theorized that these four elements were all aggregates of tiny solids. As the basic building blocks of all matter, he argued, these four elements must have perfect geometric form, namely the shapes of the five regular solids that so captivated the Greek mathematicians—the perfectly symmetric tetrahedron, cube, octahedron, icosahedron, and dodecahedron. (See Figure 2.1.)

As the lightest and sharpest of the elements, said Plato, fire must be a tetrahedron. Being the most stable, earth must consist

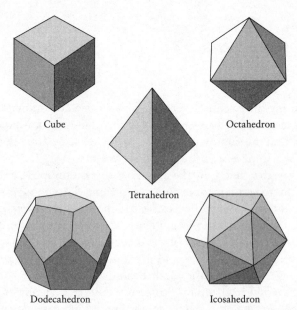

Cube   Octahedron

Tetrahedron

Dodecahedron   Icosahedron

Figure 2.1. The five regular solids of classical Greek geometry: the tetrahedron (4 triangular faces), the cube (6 square faces), the octahedron (8 triangular faces), the dodecahedron (12 pentagonal faces), and the icosahedron (20 triangular faces). For each solid, each face is a regular polygon, with all sides equal and all angles the same.

of cubes. Water, because it is the most mobile and fluid, has to be an icosahedron, the solid that rolls most easily. As to air, he observed somewhat mysteriously that " . . . air is to water as water is to earth," and concluded, even more mysteriously, that air must therefore be an octahedron. Finally, to account for the last regular solid, he proposed that the dodecahedron represented the shape of the entire universe.

Although the particulars of Plato's theory can easily be dismissed as whimsy, the philosophical assumptions behind it are exactly the same as those that drive science today. The universe is constructed in an ordered fashion that can be understood using mathematics. To Plato as to many others, God must surely have been a geometer. Or as the great Italian scientist Galileo Galilei wrote in the seventeenth century, "In order to understand the universe, you must know the language in which it is written. And that language is mathematics."

Believing the world was constructed according to mathematical principles, Plato simply took for his fundamental particles the most perfect piece of mathematics then known. That was the proof (found in Euclid's *Elements*) that there are exactly five regular solids—solid objects for each of which all the faces are identical equal-angled polygons that meet at equal angles.

As recently as the seventeenth century, the astronomer Johannes Kepler, who discovered the mathematical formula that describes the motion of the planets around the Sun, was likewise seduced by the mathematical elegance of the regular solids. There were six known planets in Kepler's time: Mercury, Venus, Earth, Mars, Jupiter, and Saturn, and a few years previously Copernicus had proposed that they all revolved in circular orbits with the Sun at the center. (Kepler would later demonstrate that the orbits were not circles, but ellipses.) Starting from Copernicus's suggestion, Kepler developed a theory to explain why there were exactly six planets and why they were at the particular distances from the sun that he and other astronomers had recently measured. There were six planets, he reasoned, because between each adjacent pair of orbits (think of the orbit as a circle going round a spherical ball in space) it must be possible to fit, snugly, an imaginary regular solid, with each solid used exactly once.

After some experimentation, he managed to find an arrangement of nested regular solids and spheres that worked: The outer sphere (on which Saturn moves) contains an inscribed cube, and in that cube is inscribed in turn the sphere for the orbit of Jupiter. In that sphere is inscribed a tetrahedron; and Mars moves on that figure's inscribed sphere. The dodecahedron inscribed in the Mars-orbit sphere has the Earth-orbit sphere as its inscribed sphere, in which the inscribed icosahedron has the Venus-orbit sphere inscribed. Finally, the octahedron inscribed in the Venus-orbit sphere has itself an inscribed sphere, on which the orbit of Mercury lies. Figure 2.2 presents Kepler's painstakingly detailed illustration of his theory.

Of course, Kepler was completely wrong. For one thing, the nested spheres and the planetary orbits did not fit together particularly accurately. Having himself been responsible for producing much of the accurate data on the planetary orbits, Kepler was certainly aware of the discrepancies, and he tried to adjust his model by taking the spheres to be of different thicknesses, though without giving any reason why the thicknesses should differ. In any case, we now know that there are not six planets but eight (nine if you count Pluto).

Still, Kepler, like Plato, was employing the same underlying philosophy that drives all present-day scientific theorizing about the universe: that it operates according to mathematical laws.

At the beginning of the twentieth century, the accepted theory of matter was the atomic theory, which viewed everything as being made up of atoms, miniature "solar systems" in which a number of electrons (the "planets") orbited a central nucleus (the "sun"). For a brief moment, God must indeed have seemed to be a geometer.

Despite the fact that the appealing picture of "solar system atom" endures to this day in the popular imagination, its life as a scientific theory was extremely short. Within twenty years, scientists had observed phenomena that forced them to accept that this model would have to be abandoned or drastically modified.

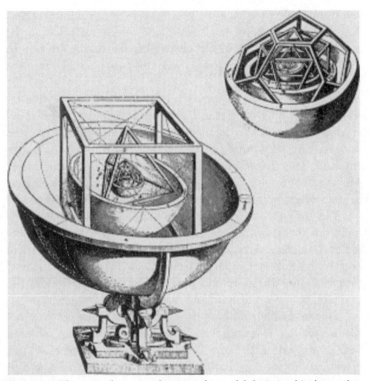

Figure 2.2. This is Kepler's own drawing of a model depicting his theory that the six known planets were situated according to a nested sequence of the five regular solids.

What they found to replace it was a far more complicated mathematical explanation known as quantum theory. As it developed in the 1920s, quantum theory embraced the principle that there is a built-in uncertainty about matter. If you were to focus attention on a single particle, such as an electron, you would find that you could not simultaneously know both its position and its momentum—the better you fixed one value, the fuzzier the other became. The electron's behavior could be described mathematically only by using probability theory, the theory that a group of European mathematicians had developed in the seventeenth century to help their wealthy patrons win at the gaming tables. Although quantum theory is nowadays widely accepted, in the early days its dependence on probability led Albert Einstein to dismiss it with the remark that "God does not play dice with the universe."

Today, the quantum-theoretic lens has been focused on matter at an ever finer scale than that of electrons, to reveal that everything in the world consists ultimately of tiny folds and ripples in space-time (the study of which requires still more new mathematics). This development has led the writer and broadcaster Margaret Wertheim to quip, "These days, God isn't a geometer, he does origami."

The common theme that links Plato, Kepler, Einstein, the quantum theorists, and present-day string theorists is the belief that an understanding of the basic stuff of the universe will be found using mathematics. Sometimes, the mathematicians' progress runs ahead of the physicists'. Today, however, the mathematicians are behind and trying to catch up.

This situation is nothing new. There have been previous occasions when scientists have put forward theories that required new mathematics. In the seventeenth century, Isaac Newton's investigations into mechanics (the motion of the planets and of moving bodies on Earth) and optics led him to develop calculus. More recently, Einstein's observation that gravity could be understood using a strange new kind of geometry developed not long before as an idle curiosity by the mathematician Bernhard Riemann led to the massive development of this geometry and its rapid incorporation into mainstream mathematics. And quantum theory itself required the development of the new branches of mathematics known as functional analysis and group representation theory.

The second of the Millennium Problems is a specific puzzle that mathematicians must solve in order to rise to the challenge laid down by the physicists. It asks for the development of a particular piece of new mathematics that, scientists believe, will help us move forward in our understanding of matter. The name the Clay Institute has given to this problem is Yang–Mills Theory and the Mass Gap Hypothesis. It is very definitely a problem of mathematics. But to understand what it means and where it comes from, we must start with the physics.

The Holy Grail of Modern Physics

When it comes to understanding the physics of the familiar everyday world—in particular force and motion—Isaac Newton pretty well got it right back in the seventeenth century. Newtonian physics allows you to predict the time it will take an object dropped from a table to reach the ground. The same physics got Apollo astronaut Neil Armstrong to the Moon and back. But we've known for a long time that Newton's physics becomes increasingly inaccurate on an astronomical scale, where the distances involved exceed (say) the diameter of the solar system, and likewise does not apply to the submicroscopic world inside the atom, where the distances get smaller still.

Early in the twentieth century, physicists developed new theories to explain these two "extreme" worlds, the world of the very large and the world of the very small. Einstein's relativity theory describes the universe on an astronomical scale, and quantum theory describes the world on a subatomic scale.

Both theories have proved extremely successful, and each is supported by an impressive catalogue of experimental and observational evidence. Moreover, although each is contrary to Newtonian physics, we can regard Newton's theory as an approximation (to the middle-sized world we experience) of either theory. On those grounds, both relativity and quantum theory would appear to have a legitimate claim to provide us with a more accurate picture of the way the universe works than Newton's theory. So which one has the better claim? The answer is neither. Relativity theory and quantum theory contradict one another. Each is very definitely wrong in precisely the domain where the other works so well.

Now, for the most part, physicists study the universe *either* on the very large scale of stars, galaxies, and beyond (astrophysicists use relativity theory) *or else* on the small scale of the atomic level and smaller (particle physicists use quantum theory). Thus, the contradictions between the two theories rarely surface. But sómetimes they do.

One place where the conflict definitely does hit you is inside a black hole, where the gravitational collapse of matter leads to a region that, in terms of its physical behavior, is simultaneously both very large and very small. Not surprisingly, we don't have any good theories that describe what goes on inside a black hole. Attempts to combine relativity theory and quantum theory lead to equations whose solutions are infinite, hence meaningless. For similar reasons, physicists have been unable to explain what went on at the moment of the Big Bang, when our universe was first created and all the matter of the universe was crammed into a tiny region.

The conflict between relativity and quantum theory also arises when physicists examine the nature of matter on the finest scale possible. According to current theories, the fundamental particles of matter that are the stuff of quantum theory—such as photons, quarks, and bosons—are really just "energy ripples in space-time" (the evocative term currently favored is "quantum foam"). Thus to study them you have to make use of Einstein's celebrated equation $E = mc^2$, which establishes—*within relativity theory*—the relationship between matter and energy.

Clearly, the fundamental conflict between relativity and quantum theory is a strong indication that neither can be the ultimate theory of matter. After all, the stars, planets, and galaxies— the objects to which relativity applies so well—are all made of the subatomic particles that quantum theory describes so beautifully. Thus if we are ever to truly understand the universe, we will have to find a single, overarching supertheory to which relativity theory and quantum theory are each approximations. But what is that theory? Physicists have been looking for it ever since Einstein, hitherto without success. Indeed, such has been the obsession of at least some of the leaders of the field that the search for the Grand Unified Theory of matter (or GUT for short) has been described as the Holy Grail of modern physics.

As we shall see shortly, the hunt boils down to finding a single framework that explains what are now believed to be the four (and only) fundamental forces of nature: electromagnetism, gravity, the strong nuclear force, and the weak nuclear force (about each of which I'll have more to say later). Currently, the electromagnetic force can be explained on the basis of Newtonian physics; Einstein's theory of general relativity explains gravity; and quantum theory provides the basis for explaining the two nuclear forces. The sought-after GUT will provide a single "superforce" of which these four fundamental forces are all special cases.

Since 1925 or so, most the the effort in searching for a GUT has been in the development of an extension of quantum theory that physicists call "quantum field theory" (QFT). The picture of matter that QFT has given us, which represents our best current knowledge of the nature of the material that makes up the universe, is generally referred to as the "standard model of particle physics."

One of the present leaders in this ongoing research is the physicist Edward Witten of the Institute for Advanced Study, in Princeton, New Jersey. So great has been Witten's influence, and so far ahead of anyone else has he often seemed, that many have compared him to Isaac Newton, who dominated physics and mathematics in the seventeenth century and whose scientific achievements stand out as remarkable even today. Witten has described QFT as "a twentieth century scientific theory that uses twenty-first century mathematics." By that, he means that much of the mathematics remains to be worked out. (It may seem that Witten is being hard on the mathematicians for being so tardy, but in fact he is being optimistic. Much of Newton's science depended on the methods of calculus, which he invented for the purpose, but which was not fully worked out as a mathematical theory until two hundred years later!)

The Yang–Mills Theory, first proposed in the 1950s, is an initial step toward such a grand unified theory. The Mass Gap Hypothesis is a particular mathematical problem arising in the

Yang–Mills framework. Together they make up the second Millennium Problem. To understand this problem and how it arose, we'll start back in the early nineteenth century, when scientists were trying to understand the properties of a newly discovered phenomenon that would come to play a major role in human life: electricity.

The Experimental Nuisance That Changed the World

Working in his laboratory one day in 1820, the Danish physicist Hans Christian Oersted noticed that a magnetic needle was deflected when an electric current was passed through a wire in its vicinity. His assistant shrugged: It was a nuisance that happened all the time, and they should simply ignore it. In an act that was to have profound consequences, Oersted did not follow his assistant's advice. Instead, he reported his observation to the Royal Danish Academy of Sciences as a scientific finding. There seemed to be, he said, a connection between magnetism and electricity.

The following year, a Frenchman called André-Marie Ampère made a similar discovery. He observed that when an electric current passes along two parallel wires, they behave like magnets. When Ampère sent a current through the two wires in the same direction, the wires attracted each other; when he directed the current in opposite directions, they repelled each other. Ten years later, in 1831, the English bookbinder Michael Faraday and the American schoolmaster Joseph Henry independently discovered the opposite effect: When they subjected a wire coil to an alternating magnetic field, an electric current was induced in the coil.

The conclusion was unavoidable. For all their seeming difference, electricity and magnetism seemed to be closely connected and somehow interchangeable.

Hearing of these observations, the great English physicist William Thomson (Lord Kelvin) wondered whether electrical force and magnetic force were different manifestations of the same underlying phenomenon. Having recently developed a mathematical theory of the motion of fluids, Thomson suggested that it might be possible to explain magnetism and electricity in a similar fashion, as two aspects of some kind of force. But a force acting in what? In the case of fluid flows, the forces causing the flow acted through the fluid itself. But what would "carry" the force giving rise to electricity and magnetism?

An old idea was that space was permeated by a continuous substance called the "ether." The ether was assumed to be uniform and at rest throughout space, a constant backdrop against which the stars and the planets moved and through which heat and light flowed. Thomson suggested that electricity and magnetism arose from some form of "force field" in the ether. A force field is not a difficult concept. You have a force field, or more simply a field, whenever there is a force acting at each point in some region of space.

For an example you need look no further than the present-day Jacuzzi. As you move your hand or your body around in the water you feel the different forces of the water coming from different directions. At any point in the water there is a force (a flow of the water) that you can feel, but the magnitude and direction of the force change from location to location and possibly from moment to moment as well. Because the forces vary from location to location, it would make no sense to speak simply of "the force of the water." Rather, a physicist or a mathematician would refer to the force *field*, the entire structured aggregate of all the forces at every point.

For another example—one directly relevant to the subject of this chapter—students in high school science classes usually view the "lines of force" in a magnetic force field by placing a sheet of cardboard on top of a magnet and sprinkling iron filings onto the card. When the card is tapped lightly, the filings align themselves into an elegant pattern of curved lines that follow the invisible lines of magnetic force. (See Figure 2.3.) The iron filings provide a picture of the force field—more precisely, they illustrate a two-dimensional cross section of the three-dimensional field.

In general, you have a force field whenever there is a specific force at each point in a region, with both the magnitude and the direction of the force varying with position—generally in a con-

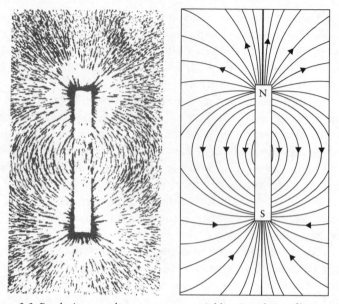

Figure 2.3. By placing a card over a magnet, sprinkling it with iron filings, and then gently tapping the card, the filings align themselves with the otherwise invisible lines of magnetic force, providing a map of the magnetic force field generated by the magnet.

tinuous, unbroken fashion. If you were to move about in such a field, the force you would be subjected to would vary in both magnitude and direction. In many force fields, the force at each point also varies with time. Mathematicians study force fields by formulating equations that specify the magnitude and direction of the force at each point (and possibly at each moment as well). The mathematical structure that arises in this way is called a vector field. (A vector is a quantity that has both magnitude and direction, such as velocity or force.)

Maxwell Sees the Light

Around 1850, the Scottish mathematician James Clerk Maxwell decided to investigate Thomson's suggestion that electrical force and magnetic force were two aspects of a single phenomenon. Some fifteen years later he had his answer, together with a name for the unifying force: "electromagnetism." He published his results in 1865 in a book titled *A Dynamical Theory of the Electromagnetic Field*.

The formulas Maxwell wrote down to describe an electromagnetic field are known today simply as Maxwell's equations. Like the Navier–Stokes equations that we shall meet in Chapter 4, they are partial differential equations. They describe the connection between an electrical field **E** and a magnetic field **B**, where **E** is a vector function that for each point and each moment of time gives an electric current (a vector) at that point, and **B** is a vector field that for each point and each moment of time gives a magnetic force (also a vector) at that point. (Vectors are often denoted by boldface symbols, to indicate that they have both magnitude and direction.)

By finding a common framework that united what had previously been regarded as two separate forces (electricity and magnetism), Maxwell began the quest to unify the forces of nature that has guided fundamental physics ever since.

Maxwell's equations imply that if an electric current is allowed to fluctuate back and forth in a conductor, the electromagnetic field, which alternates in time with the current, will tear itself free from the conductor and flow into space in the form of an electromagnetic wave. The frequency of the wave will be the same as that of the current that causes it. (Maxwell referred to the electromagnetic flow as a *wave* only because the equations that described the flow were similar to the ones that describe waves in fluids. But as we shall see, the question of whether electromagnetic radiation really is a "wave" turned out to be both tricky and deep.)

Wave or not, Maxwell was able to calculate the speed of the electromagnetic flow that breaks free, finding that it was about 300,000 kilometers per second (about 186,000 miles per second). There was something familiar about this figure. It was very close to the speed of light. Indeed, it was so close that Maxwell suspected that the two speeds might actually be the same, leading him to speculate that light was simply electromagnetic radiation of particular frequencies. Though this suggestion did not meet with immediate acceptance by all scientists, today we know he was right. Light is indeed a form of electromagnetic radiation.

Electromagnetic radiation is generally measured in terms of the frequency of the wave. The waves having the lowest frequency are the radio waves used to transmit radio and television signals. At higher frequencies come microwaves and infrared waves, which are invisible but transmit heat. Then comes light, the visible portion of the electromagnetic spectrum, with red at the lower frequency end and violet at the higher frequency end, and the familiar colors of the rainbow – orange, yellow, green, blue, and indigo – arranged in order in between. Radiation with a frequency slightly higher than violet it known as ultraviolet, which, though not visible to the human eye, will blacken a photographic film and can be seen using special equipment. Beyond the ultraviolet region we find the invisible X-rays, capable not only of blackening a photographic film but also of penetrating human flesh, a combination of features that has led to their widespread use in medicine. Finally, right at the upper end of the spectrum are the gamma rays emitted by radioactive substances as they decay, a form of electromagnetic radiation that in recent years has also been used in medicine.

One question that Maxwell's equations did not answer was what exactly the medium was through which electromagnetic waves traveled. A number of scientists attempted to detect it, or more precisely the motion of the Earth through it, including a famous experiment performed by Albert Michelson and Edward Morley in 1887, but all attempts failed. Attention shifted dramatically, however, with the arrival on the scientific scene of a young man working in a patent office in Switzerland.

Einstein and the Theory of Special Relativity

In the year 2000, *Time* Magazine took it upon itself to name the most important person of the twentieth century. The winner was a German-born individual who began his career as a lowly Swiss patent officer and ended it as the most famous scientist in the world: Albert Einstein.

Born in Ulm in 1879, Einstein spent most of his childhood in Munich, where he received his schooling. In 1896, he relinquished his German citizenship because of his dislike of German militarism, and was stateless until he obtained Swiss citizenship in 1901. By then, he had moved to Zurich and had graduated from the Swiss Polytechnical Institute. In January 1902, having failed to gain a university position, he took a job as Technical Expert, Third Class, in the Swiss Patent Office in Bern.

Three years later, in 1905, he developed his famous theory of special relativity, a scientific breakthrough that would make him world famous within a few years. In 1909 he resigned his position at the Patent Office to take up an appointment of Extraordinary Professor of Physics at the University of Zurich. Ten years later his Theory of General Relativity was experimentally confirmed, and from then until his death in 1955 he enjoyed a measure of celebrity unmatched by any other scientist.

To get an initial understanding of special relativity, imagine you are in an airplane, flying at night with the window blinds down, so you cannot see outside. Assuming there is no turbulence, you are not aware of the fact that the airplane is in motion. You get up from your seat and walk around. The flight attendant pours you a cup of coffee. Everything seems normal, just as if you were at rest on the ground. And yet you are hurtling through the air at 500 miles an hour. How is it that when you get up from your seat, you are not thrown to the back of the plane? How is it that the poured coffee does not stream back into your chest?

The answer is that the motion of you and the coffee are relative to the motion of the airplane. The interior of the airplane provides a "stationary" background—what physicists call a frame of reference—relative to which you and the coffee move. From the point of view of you or anyone else inside the airplane, everything behaves exactly as it would if the airplane were at rest on the ground. It is only when you open the window blinds and look outside, seeing lights beneath you moving steadily backwards, that you can detect the airplane's motion. You do that because you are able to compare two frames of reference: the airplane and the ground.

This example shows that motion is relative: One thing moves relative to another. What we experience as absolute motion is motion relative to the frame of reference we are in—and are aware of. But is there a "preferred" frame of reference—nature's own frame of reference if you like (such as the postulated ether)? For Aristotle, the Earth was at rest, and hence motion relative to the Earth was "absolute" motion. Copernicus believed that all motion is relative. Newton believed in a fixed "space" against which everything was either absolutely stationary or else in absolute motion. Einstein went a step further. He abandoned the idea of a fixed framework in space—and with it the concept of a stationary ether—and simply declared that all motion is relative. According to Einstein, there is no preferred frame of reference. This is Einstein's principle of special relativity.

A surprising consequence of this principle is that electromagnetic radiation has a very special property. Whatever your frame of reference, when you measure the speed of light or any other form of electromagnetic radiation, you will always get the same answer. For Einstein, the one absolute was not some material substance through which electromagnetic waves traveled—the ether—but the speed of electromagnetic waves.

By taking the speed of light as being the same in all frames of reference, Einstein was able to resolve another troublesome question: What does it mean to say that two events happen at the same time? This becomes problematic when the events occur a great distance apart. Einstein solved the problem of simultaneity by declaring that time was not absolute, but depended on the frame of reference in which it was measured. Clearly, the mathematical analysis of the universe was starting to lead scientists into some decidedly counterintuitive territory.

Gravity: The Theory of General Relativity

For all its power, Einstein's theory of special relativity applied only when two or more frames of reference moved relative to each other at constant velocity. Moreover, although the theory described the nature of space and time, it said nothing about those other basic constituents of the universe: mass and gravity. In 1915, Einstein found a way to extend his theory of relativity to take account of both. He called his new theory general relativity.

The basis for the new theory was the principle of general relativity: that all phenomena occur the same way in all frames of reference, whether in accelerated motion or not. In general relativity, a natural process affected by gravity would occur if there were no gravity and the whole system were in accelerated motion. Gravity and acceleration are interchangable.

For a particular example of the general relativity principle, consider again the scenario where you are a passenger on a night-time airline flight with all the window shades down. If the aircraft makes a sudden acceleration, you will feel a force pushing you toward the back of the airplane. If you happen to be standing in the aisle at the time, you might find yourself flung toward the rear of the plane. Likewise, when the plane decelerates rapidly (as it does on landing), you are subjected to a force toward the front of the plane. In both cases, you perceive acceleration or deceleration as a force. Since you are unable to see outside, you cannot observe that the airplane is speeding up or slowing down. If you did not know any better, you might be inclined to try to explain your sudden motion toward the back or the front of the plane as caused by some mysterious force. You might even decide to call that force "gravity."

Confirmation of Einstein's bold theory came in a dramatic fashion. An initially surprising consequence of general relativity is that light should behave as if it had mass. In particular, light waves should be subject to gravity. Therefore, if a light wave passes close by a large mass, such as a star, the gravitational field of the mass should cause the light wave to deflect. In 1919, the British astronomer Sir Arthur Eddington made an accurate observation of the planet Mercury during a total solar eclipse, finding that its position in the sky was "wrong." The explanation was that the sun caused the light from the planet to bend, resulting in the apparent shift in its position. The amount of shift agreed exactly with Einstein's theory, and so the conclusion was inescapable. Although Newton's theory of gravity and planetary motion was sufficiently accurate for most everyday purposes, such as calculating lunar calendars and drawing up tide tables, when the time came to do precise astronomical science, Einstein's theory was much more accurate than Newton's.

The theory of general relativity did more than provide accurate predictions of astronomical gravitational effects. It also provided an explanation of the very nature of gravity. Surprisingly, it was a geometric explanation. According to Einstein, any solid object distorts space-time, causing it to curve. The degree of curving—what mathematicians call the curvature—is greatest immediately surrounding the object, and decreases the further away you go. This curvature is what causes two objects to exert an attractive force on each other. In other words, what we call gravity is, in Einstein's framework, simply (a manifestation of) the curvature of space-time caused by the presence of an object. The bigger and more massive the object, the greater the distortion it creates in the surrounding space-time, and thus the greater the gravitational force it exerts. The further away from the object you are, the less the distortion, and hence the gravitational force decreases as you move away from its source.

For all its importance, Einstein's theory of relativity did not earn him a Nobel Prize. He did get one, in 1921, but it was awarded for an early contribution to the very theory that would come to vie with relativity as a theory of the universe: quantum theory.

Quantum Theory: What's the Matter?

While general relativity describes the geometric structure of the universe and shows how matter affects and is affected by that structure, it does not answer the question, What exactly *is* matter? To resolve that problem physicists had to turn to another theory: quantum theory.

As we observed earlier, at the start of the twentieth century, the accepted view was that matter was made up of tiny solar-system-like atoms, each consisting of a heavy nucleus around which orbited one or more much lighter electrons. The nucleus of the atom was itself supposed to be made up of two kinds of fundamental particles: protons and neutrons. Each proton carries a unit positive charge, each electron a unit negative charge, and it is the electromagnetic attraction between these positive and negative charges that holds the electrons in their orbits around the nucleus. (This picture is still a useful one—for example chemists rely on it all the time—but we now know that it's overly simple.)

An obvious question was, What exactly are those fundamental particles—the electrons, the protons, and the neutrons—of which atoms are built? That was the issue addressed by the physicists Niels Bohr, Werner Heisenberg, and Erwin Schrödinger around 1920. In trying to provide an answer, they had to take account of some strange experimental results, one being that light seemed to behave like a continuous wave in some circumstances and a stream of discrete particles in others. The answer they came up with—quantum theory—certainly resolved all of the puzzles, but in so doing it introduced a feature that seemed antithetical to the basic scientific idea of a universe that operated by cause and effect. Quantum theory incorporated an element of randomness in a fundamental way.

The first major step toward quantum theory came from attempts in the late nineteenth century to resolve a puzzle created by Maxwell's electromagnetic theory. According to that theory, the total energy generated in an enclosed "black oven" (a sealed, lightproof metal box whose walls are heated to red hot) should be infinite, since the heated walls of the oven would emit electromagnetic radiation of all possible frequencies—of which there are infinitely many. In 1900, the German physicist Max Planck suggested a resolution to the puzzle. He posited that energy comes in discrete "clumps", which he called quanta, which cannot be subdivided. The number of quanta carried by a given electromagnetic wave is proportional to the frequency of the wave. The higher the frequency, the more quanta. This at once resolves the problem of the infinite energy, since beyond a certain frequency, the energy of any wave will come in too great an amount to be part of the aggregate energy inside the oven, and thus that wave will not contribute to the total, which is therefore simply a finite sum.

Planck calculated the proportion between the energy and the frequency of a wave—what is now known as Planck's constant. Because this number is so small (its value is 6.626×10^{-34} joule-seconds, where the joule is a measure of work done), the discrete nature of energy is not apparent in everyday measurement. But when Planck computed the total energy in a black oven, his answer agreed extremely well with experimental results, adding enormous credibility to his theory. He was awarded the 1918 Nobel Prize in Physics for his work. Still, Planck was not able to answer the obvious question: Why *should* energy come in clumps?

The answer to that puzzle was provided by Einstein in 1905, as a consequence of his explanation for another strange physical phenomenon known as the photoelectric effect. (This was the result that led to Einstein's Nobel Prize, in 1921.) In 1887, the German physicist Heinrich Hertz had noticed that when electromagnetic radiation (light) falls upon certain metals, they emit electrons. The intriguing aspect of this phenomenon was that increasing the intensity of the light resulted in more electrons being emitted, but the electrons did not have any greater energy, whereas increasing the frequency of the light (i.e., changing its color toward the violet) led to the emission of electrons with greater energy.

To explain this behavior, Einstein proposed that a light wave consists of discrete packets of energy—subsequently christened photons—with the energy of each photon proportional to the frequency of the wave (with Planck's constant as the ratio). When a photon of sufficient energy strikes the metal, it can dislodge an electron. The number of electrons dislodged depends on the number of energetic photons, hence on the intensity of the light. The energy of each released electron depends on the energy of the photon that strikes it (hence on the frequency, or color, of the light).

Einstein's new picture of light provided an explanation for Planck's suggestion that the energy in an electromagnetic wave comes in discrete packets. Those packets of energy—Planck's quanta—are the photons that, according to Einstein, make up the wave.

Now, the idea that light consists of a stream of particles was not itself new. It was proposed by Newton in the seventeenth century. What was novel about Einstein's suggestion was that the photons constituted not a stream but a wave.

But what exactly does it mean to say that particles constitute a wave? (I promised that this question would come up again.) Here we have a conundrum that, it is safe to say, no human has ever come fully to terms with. As the late Richard Feynman, one of the leading pioneers in the development of modern quantum theory, once remarked:

> There was a time when the newspapers said that only twelve men understood the theory of relativity. I do not believe there ever was such a time. . . . After people read Einstein's paper a lot of people understood the theory of relativity in one way or other, certainly more than twelve. On the other hand I think I can safely say that nobody understands quantum mechanics.[1]

When it comes to quantum mechanics, physicists have to abandon their intuitions and rely on the mathematics to tell them what's going on. Mathematics began as one of several ways of understanding the world, but with the arrival of quantum theory, mathematics became *our only way to understand*.

For instance, is light a wave—a continuous ripple on the ether—or a stream of discrete particles hurtling through space?

---

1. Richard Feynman, *The Character of Physical Law*, Cambridge, MA: MIT Press, 1965, p. 129.

When light is studied in the laboratory, sometimes it behaves like one and sometimes like the other. Faced with such behavior, the only way scientists could make progress in their investigation of electromagnetic radiation was to fall back on the mathematics. And that mathematics led them into very strange territory, completely unlike anything we experience in our everyday lives. According to the mathematics, the wavelike behavior of light was a consequence of an inherent uncertainty about matter itself (including photons).

Quantum theory tells us that we have to abandon the familiar image of a particle (such as an electron or a photon) as having, at any time, a definite position or a definite velocity. The most precise information you can get about the position and velocity of a particular particle is a probability distribution that tells you the particle's *tendency* to be in a given state—the *likelihood* of where it is and what its direction and magnitude of motion are. In place of the familiar picture of a particle as occupying a "point" in space and having a particular velocity, we have to imagine a sort of cloud, a region of space (without a definite boundary) within which the particle resides. This is not simply a matter of needing more precise measurements. The uncertainty is a fundamental aspect of physical reality.[2]

At this point, let's pause for a moment and look again at the seemingly simple solar-system picture of the atom that quantum theory usurped.

The Forces of Nature

We know what keeps the electrons in their orbits around the nucleus: the electromagnetic force that causes oppositely charged particles to attract each other. But what holds the protons together in the nucleus? After all, like charges repel. Why doesn't

---

2. Mathematically, the particle is described by a wave function Ψ, which to each point x in space-time associates a vector $\Psi(x)$, whose magnitude represents the amplitude of the oscillation and whose direction gives the phase. The square of the length of $\Psi(x)$ gives the probability that the particle is close to the point x in space-time.

the nucleus simply explode? There must be some force—and a massive one at that—that holds the nucleus together. Physicists call it the strong nuclear force, or the strong interaction. The strong interaction has to be strong in order to hold together the protons in a single nucleus. On the other hand, it can act only over a very short distance, on the order of the nucleus itself, since it does not bring together the protons in nuclei of two distinct atoms. (If it did, everything in the universe would implode, including ourselves.)

The strong nuclear force is assumed to be a fundamental force of nature, along with gravity and the electromagnetic force. Just as the photon is the smallest constituent of an electromagnetic field—the carrier of electromagnetic force—so too the strong nuclear force is carried by fundamental particles called gluons.

In fact, physicists believe there is a fourth fundamental force. In order to explain nuclear decay in radioactive elements such as uranium, there must be a second kind of nuclear force that can cause protons to fly apart: the weak nuclear force. The particles that carry this force are called (weak gauge) bosons.

Both of the nuclear forces are very short-ranged—their effects are felt only within the nucleus itself. The strong force is an attractive one; the weak force is repellent. But whereas the strong nuclear force is extremely powerful within the range of the nucleus, the weak nuclear force is, as its name suggests, far less so. Only when the nucleus contains sufficient protons is their combined weak force sufficient to eject one or more protons from the nucleus, leading to nuclear decay.

At this point, the skeptic might be inclined to think that there are still further natural forces waiting to be discovered. While this cannot be completely ruled out, physicists think this is not the case. They believe that gravity, the electromagnetic force, and the strong and weak nuclear forces are all there are. A complete theory of matter will have to incorporate all four forces.

What is required, then, is a quantum-theoretic account of fields—a so-called quantum field theory, or QFT—that includes descriptions of the four fundamental force fields of nature. Traditional accounts of quantum theory, such as the brief one I gave earlier, are generally restricted to the theory as originally conceived in the 1920s, often referred to as quantum mechanics. That theory is couched in terms of particles. In quantum *mechanics*, the behavior of a particle can be described by a single equation that is well understood: the Schrödinger equation

$$ i\hbar \frac{\partial \Psi}{\partial t} = H\Psi $$

where Ψ is the wave function, H is the Hamiltonian, $\hbar$ is Planck's constant, and $i = \sqrt{-1}$. Don't worry if none of this means anything to you. I'm just including it to indicate that a single partial differential equation pretty well captures everything physicists needed to know. (Compare it with the vector form of the Navier–Stokes equation in Chapter 4.) The crucial thing to be aware of is that mathematicians have little trouble handling such equations. The early development of quantum mechanics might have led to results the human mind could not fully grasp, but by and large the mathematics was fairly straightforward.

In quantum field theory, by contrast, matter is assumed to be a kind of field. The classical picture of material particles occupying a surrounding space is no longer appropriate, even when modified to allow for inherent uncertainties in any particular position and state. Instead, the theory postulates an underlying quantum field, a fundamental continuous medium present everywhere in space. What classical physics refers to as "particles" are simply local densities—concentrations of energy, or "ripples"—in the quantum field. In QFT, the mathematics becomes much more difficult. Indeed, much of it has not yet been worked out! The discovery, in 1973, of a property called asymptotic freedom (of quantum nonabelian gauge theory), about which I'll say a little more in due course, has helped physicists to understand what kind of results the mathematics *should* lead to (among them the Mass Gap Hypothesis), but so far no one has any idea how to prove any of those conjectured results.

Notice, by the way, that Maxwell's electromagnetic theory is an example of a classical (i.e., nonquantum) *field theory*, since it treats electromagnetism not as a stream of particles—photons— but as a field. Thus, there's nothing inherently difficult in using mathematics to describe fields. It's the attempt to merge the ideas of quantum theory and field theory that causes all the difficulty.

This view of matter as a kind of property of a space-time field led to some surprising consequences. One was the prediction of antimatter: For every particle there has to be a corresponding antiparticle having the same mass but opposite electrical charge. Since matter and antimatter cannot normally exist together—if a particle comes into contact with its antiparticle, the two immediately annihilate one another—it was hardly surprising that physicists had not previously discovered antimatter in the laboratory. Thus when quantum theory pioneer Paul Dirac predicted the existence of antimatter in 1931, his suggestion met with some skepticism. But it was not long before the antiparticle of the electron—called the positron—was observed in cosmic rays, making Dirac's conclusion one of the first (of many) predictive successes of quantum field theory.

Interestingly, for all that quantum field theory seems to fly in the face of human intuition, the key mathematical idea that started physicists on the path toward the theory, and which they believe will eventually lead them to the desired GUT, is one that lies at the heart of our sense of beauty: the concept of symmetry.

Nature's Symmetries

In everyday language, an object is said to be symmetric if it has some form of balance. For example, we say that a human face is symmetric because the left-hand side looks like the right, only reversed. We say a flower or a snowflake is symmetric because, viewed from directly above, each part looks like the part directly opposite. (See Figure 2.4.)

Mathematically, an object is said to be symmetric *with respect to some motion or transformation* if that motion or trans-

Figure 2.4. Symmetry. A snowflake exhibits reflective symmetry about the center.

formation leaves it looking exactly as it did in the first place. Thus, the human face is symmetric for left–right reversal because it looks (almost) the same if we reverse the left and right sides. This is why our photograph looks almost the same as our reflection in a mirror. A flower or a snowflake is symmetric for reflection in the center because it looks the same if we swap every point with the one diametrically opposite. A square is symmetric for 90° rotations about the center because it looks the same if we rotate it about its central point through a right angle.

Mathematicians in the nineteenth century discovered that the collection of all symmetries of a given object (i.e., the collection of all transformations that leave the object looking exactly the same as at the start) has some interesting structural properties, independent of the object concerned. In particular, it has an "arithmetic"—you can "add together" two symmetries of an object to give a third symmetry, and this "addition" has some of the familiar properties of the ordinary addition of numbers.

Mathematicians called these new kinds of arithmetics "groups," and over the years developed an important new branch of mathematics—group theory—devoted to their study. (See the appendix to this chapter for a brief introduction to group theory. We will meet it again in other chapters of the book.) Group theory now forms a central part of the college education of every mathematician. As we shall see momentarily, it is also an important tool in physics.

The set of all the symmetries of a given object is called the symmetry group for that object. Knowledge of the arithmetical properties of the symmetry group can tell you a lot about the object—its shape and various other properties. Crucially, the objects to which group theory can be applied in this way don't have to be physical objects such as faces, snowflakes, or flowers. They can be abstract mathematical objects, equations, or force fields.

Early in the twentieth century, physicists began to realize that many of their conservation laws arise from symmetries in the structure of the universe. For example, many physical properties are invariant under translation and rotation. The results of an experiment do not depend on where the laboratory is situated or the direction the equipment faces. This invariance implies the classical physical laws of conservation of momentum and angular momentum.

The German mathematician Emmy Noether proved that this is true in general: Every conservation law can be regarded as the result of some symmetry. Thus, every conservation law has an associated group—the corresponding symmetry group—which describes the relevant symmetries at each point in space-time. For example, the classical law of conservation of electric charge has an associated symmetry group. So too do the quantum physicist's laws of conservation of such properties as "strangeness" and "spin."

A Tour de Force

In 1918, the mathematician Hermann Weyl set out to use the notion of symmetry in an effort to unify special relativity and electromagnetism. His idea was to capitalize on the fact that the electromagnetic field at any point possesses certain mathematical symmetries that leave the equations unchanged. For instance, Maxwell's equations are invariant with respect to a change of scale, and this is a form of symmetry. To make use of this fact, Weyl regarded an electromagnetic field as a distortion of the relativistic length that arises when you travel around a closed curve. To do this mathematically, he had to assign a symmetry group to each point of four-dimensional space-time.

Weyl's basic idea was good, but his approach did not completely work. With the emergence of quantum theory, with its emphasis on the wave function, it became clear what the problem was. It wasn't scale that was important in Maxwell's equations but phase. Weyl had been working with the wrong kind of symmetry, and hence the wrong symmetry group! The crucial symmetry of the electromagnetic field, it transpired, was what is now known as "gauge symmetry", which means that the field equations keep their form even if the electromagnetic potentials are multiplied by certain quantum-mechanical phase factors, or "gauges". (In everyday life, a gauge is, of course, a measuring device. Just as you can get a sense of the force field in a Jacuzzi by moving your hand around in the water, so too can you—at least in principle—get a picture of an electromagnetic force field by moving some form of "measuring gauge" around in the field.) This was the birth of the hugely important new subject of "gauge theory", in which the symmetry group assigned to each point in space-time was called a gauge group.

When he emphasized scale, the symmetry group Weyl had been working with was multiplication on the positive real numbers. After he changed his focus from scale to phase, the important group for Maxwell's equations became the "one-dimensional unitary group" $U(1)$, which can be viewed as the set of rotations of the plane.

Following Weyl, physicists were able to recast Maxwell's theory as a gauge theory. Their strategy for extending Maxwell's theory to a quantum field theory incorporating one or more of the nuclear forces (and maybe even gravity) was then to replace the gauge group U(1) with a more complicated symmetry group so that the resulting field theory could first of all be a quantum field and secondly could incorporate the fundamental force fields. They achieved this extension (except for gravity) in several Nobel Prize–winning stages.

The first step occurred during the 1940s, when the physicists Richard Feynman, Julian Schwinger, Sin-Itiro Tomonaga, and others developed an extremely powerful and remarkably accurate quantum description of electromagnetism. Their theory is known as quantum electrodynamics, or QED for short. Essentially, it is a quantum-theoretic version of Maxwell's theory. It is widely regarded as the most precise scientific theory ever developed, with theoretical calculations that turn out correct to as many as eleven decimal places when measured in the laboratory—a precision unequalled in any other area of science. Feynman, Schwinger, and Tomonaga were awarded the Nobel Prize for their work in 1965.

The next key step was the formulation, in 1954, of a quantum-theoretic analogue of Maxwell's equations by the physicists Chen-Ning Yang and Robert Mills. In a brilliant move that won them the 1957 Nobel Prize in Physics, Yang and Mills replaced the group U(1) by what is known as a "compact Lie group"—a set of rigid motions of a complex multidimensional space. Whereas Maxwell's equations are entirely classical, that is, non-quantum-theoretic, the Yang–Mills equations come in two flavors: classical and quantum-theoretic. Thus Yang–Mills theory allowed for the development of a quantum field treatment of matter that extends QED.

Using the Yang–Mills equations requires much more complicated mathematics than Maxwell's theory. In particular, the group U(1) associated with Maxwell's equations is "abelian" (i.e., commutative—any two successive rotations of the plane may be performed in any order, without changing the result). But that is not true for the group Yang and Mills used. Theirs was a "nonabelian" gauge theory. The loss of commutativity makes the mathematics considerably trickier.[3]

Following the development of Yang–Mills theory, physicists set about trying to use non-abelian gauge theory to find their long sought-after grand unified theory. The general idea, remember, was to find the right gauge group to enable them to capture the two nuclear forces and—possibly hardest of all—gravity.

One problem with using a quantum version of Yang–Mills theory to unify electromagnetism with the weak or strong forces was that classical (i.e., nonquantum) versions of the Yang–Mills equations describe zero-mass waves that propagate at the speed of light. (In this regard, the Yang–Mills equations are like Maxwell's equations.) In quantum mechanics, however, every particle can be considered as a special kind of wave, and thus this "massless" feature presented a major sticking point. The nuclear forces were known to be carried by particles of *nonzero* mass.

For the weak force, this difficulty was overcome in 1967 by Sheldon Glashow, Abdus Salem, and Steven Weinberg, using a gauge theory having the symmetry group that goes by the technical name SU(2) × U(1). The theory they developed is called the electroweak theory. They avoided masslessness by introducing an extra force, the Higgs field. Current experiments aimed at detecting the Higgs boson, which carries the Higgs field, are attempts to confirm the electroweak theory.

The electroweak theory not only captured both electromagnetism and the weak nuclear force, it showed that at sufficiently high energy, such as existed in the first few moments after the Big Bang, the two forces merge into one, which the three physicists called the electroweak force. (The process whereby the electroweak force separated out into two seemingly different forces is called symmetry breaking.) Glashow, Salem, and Weinberg were awarded the Nobel Prize in 1979 for this achievement.

The next step was taken by Howard Georgi and Glashow a decade later. Using a larger group known as SU(3) × U(2),

---

3. Symbolically, Yang and Mills replaced each component of the electric and magnetic potential vectors in Maxwell's equations by a matrix.

they managed to incorporate the strong force as well, calling this new theory quantum chromodynamics (QCD). On this occasion, they overcame the obstacle of masslessness not by adding additional fields but by discovering a remarkable property of quantum Yang–Mills theory itself, a property known as asymptotic freedom. This says, roughly, that very short-range effects are nearly classical, with quantum effects showing up only at a longer range.

A major triumph for QCD was the prediction of the existence of quarks, fundamental particles with spin $\frac{1}{2}$, somewhat analogous to the electron, that combine together to form protons, neutrons, and other previously known particles.

Beyond QCD, a complete GUT incorporating gravity has proved elusive. In recent years, physicists have looked to other theories, more general than gauge theory, to complete this last step. By far the most common is string theory, of which the leading developer is the Princeton physicist Edward Witten. In string theory, the fundamental objects are no longer particles but tiny open or closed strings. To make this work, physicists have had to learn to adapt to working in a space-time manifold of at least ten dimensions. (If there are fewer dimensions, there is not enough freedom for the strings to have the required properties.)

The Second Millennium Prize Problem

The second of the seven Millennium Prize Problems arises from the developments that led to QCD. As with QED, many predictions of QCD have been confirmed experimentally to an accuracy unparalleled in science, so physicists are confident that they are on the right track. But our mathematical understanding of the theory remains far from complete. For instance, no one has been able to solve the Yang–Mills equations (in the sense of writing down a formula that gives a general solution to the equations), let alone any of their subsequent generalizations. Instead, physicists use the equations to formulate rules for calculating various key numbers in an "approximate" way. (I use quotation marks here because of the incredible accuracy of those calculations.)

When you think about it for a moment, it seems incredible. The most accurate scientific theory the world has ever seen is built upon equations that no one can solve. The Millennium Problem of Yang–Mills Theory challenges the mathematical community to address this issue, first by finding a solution to the Yang–Mills equations, and second by establishing a technical property of the solution called the Mass Gap Hypothesis. This second part of the problem will ensure that the mathematics remains consistent with computer simulations and observations made by physicists in the laboratory.

Specifically, experiments, computer simulations, and some fairly crude theoretical calculations led physicists to believe that at the quantum level, QCD has the following two key features (which it must have if it is to successfully explain the strong force):

- There must be a "mass gap"; i.e., there is a nonzero minimum energy level for excitations of the vacuum (no massless particle waves);
- It must have "quark confinement"; i.e., all particle states have SU(3) symmetry, even though more general fields than particles may not.

The mass gap property explains why the strong force has such a short range; quark confinement explains why individual quarks cannot be observed. These quantum-theoretic properties are very different from the behavior of the classical theory.

To date, no one has been able to prove either of the above two properties mathematically. The second Millennium Prize Problem includes a precisely formulated, mathematical version of the first of the two—the Mass Gap Hypothesis. Specifically, the second Millennium Problem is to prove that:

> For any compact, simple gauge group, the quantum Yang–Mills equations in four-dimensional Euclidean space have a solution that predicts a mass gap.

(A group is simple if it cannot be built up from smaller groups in a certain algebraic manner.) The solution to this one problem would represent a major breakthrough not only in theoretical physics but also in the larger quest to develop quantum field theory as a *mathematical* (and not simply physical) theory.

According to Arthur Jaffe of Harvard University, an expert in the mathematics of quantum field theory and the current director of the Clay Institute, the problem of Yang–Mills Theory and the Mass Gap Hypothesis was chosen as a Millennium Problem because its solution would mark the beginning of a major new area of mathematics having deep and profound connections to our current understanding of the universe. Edward Witten, accepting that the Millennium Problems should include at least one that comes from modern physics, justifies the choice made by the Clay Institute with these words:

> What mathematical problem best embodies the challenge of understanding quantum field theory? We want a problem that is (i) central in physics; (ii) important mathematically; and (iii) representative of the difficulties of QFT. To me, the outstanding problem with these features is this one: *Prove the existence and mass gap of quantum Yang–Mills theory on* Re^4, *with gauge group a compact, simple, nonabelian Lie group G*.[4]

Witten sees this Millennium Problem as a major challenge for humanity. Speaking in the summer of the Millennium Year 2000, he said, "Understanding natural science has been, historically, an important source of mathematical inspiration. So it is frustrating that, at the outset of the new century, the main framework used by physicists for describing the laws of nature is not accessible mathematically."[5]

For Witten, the two parts of the problem have very different significance. Finding a general solution to the Yang–Mills

---

4. Edward Witten, *Physical Law and the Quest for Mathematical Understanding*, paper presented at the University of California at Los Angeles, August 2000.
5. Ibid.

equations, he says, "would essentially mean making sense of the standard model of particle physics." As such, it would be a major achievement within mathematics—"a milestone in coming to grips mathematically with twentieth century theoretical physics."[6] But he feels this solution would not have much significance to physicists, who, in their own terms, already know that the equations work and why.

A proof of the Mass Gap Hypothesis, on the other hand, would be of enormous significance in both mathematics and physics, since the proof "would shed light on a fundamental aspect of nature that physicists still do not properly understand."[7]

So the stakes are high indeed. But what are your chances, reader, of being able to solve the second Millennium Problem? Not good, apparently. According to Witten, the problem is "a worthy 21st century challenge, but it is too hard for now."[8]

---

6. Ibid.
7. Ibid.
8. Ibid.

APPENDIX

GROUP THEORY:
THE MATHEMATICS OF SYMMETRY

The mathematical objects known as groups arose when mathematicians tried to come to grips with symmetry. In ordinary language, we say an object (say a vase or a face) is symmetric if it looks the same from different sides or from different angles, or when it is reflected in a mirror. How can we make this precise? What exactly do we mean by that phrase "looking the same from different angles"? Well, imagine you have some object in front of you, and that the object is rotated about some line or point. Does the object look the same after the manipulation as it did before? If it does, we shall say the object is "symmetric" for that particular manipulation.

For example, if we take a circle and rotate it about its center through any angle we please, the resulting figure looks exactly the same as when it started. (See Figure 2.5.) We say that the circle is symmetric for any rotation about its center. Of course, unless the rotation is through a full 360° (or a multiple of 360°), any point on the circle will end up in a different location. The circle will have moved. But even though individual points have moved, the figure looks exactly the same.

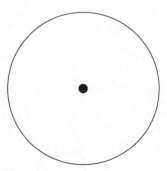

Figure 2.5. The circle looks the same after any rotation about its center.

A circle is symmetric not only for any rotation about its center, but also for a reflection in any diameter. Reflection here means swapping each point of the figure with the one directly opposite with respect to the chosen diameter. For example, with a clock face, reflection in the vertical diameter swaps the point at 9 o'clock with the point at 3 o'clock, the point at 10 o'clock with the point at 2 o'clock, and so on. (See Figure 2.6.)

The circle is unusual in that it has many symmetries, infinitely many, in fact. A square, on the other hand, has less symmetry than a circle. If we rotate a square through 90° or through 180° in either direction, it looks the same. But if we rotate it through 45°, it looks different—we see a diamond shape. Other manipulations of a square that leave it looking the same are reflecting it about either of the two lines through the center point, parallel to an edge. Or we can reflect the square in either of the two diagonals. Figure 2.7 shows that each of these manipulations moves specific points of the square to other positions, but as with the circle, the figure we end with looks exactly the same—in position, shape, and orientation—as it did before.

We have now moved from an everyday idea of "symmetry" to the more precise notion of symmetry with respect to a particular manipulation of the object. The greater the number of manipulations that leave a figure or object looking the same (in

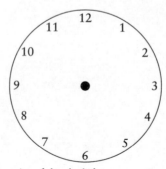

Figure 2.6. The symmetries of the clock face.

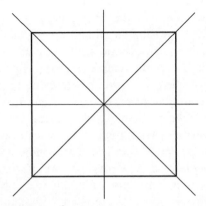

Figure 2.7. The symmetries of a square.

position, shape, and orientation), the more "symmetric" it is in the everyday sense.

Since we want to apply our concept of symmetry to things other than geometric figures or physical objects, we shall begin to use the word "transformation" rather than manipulation. A transformation takes a given object (which may be an abstract object) and transforms it into something else. The transformation might simply be translation (moving the object to another position without rotating it), or it could be rotation (about a point for a two-dimensional figure, about a line for a three-dimensional object) or reflection (in a line for a two-dimensional figure, in a plane for a three-dimensional object). Or it could be something that is not generally possible for a physical object, such as stretching or shrinking.

The key to the mathematical study of symmetry is to look at transformations of objects rather than the objects themselves.

To a mathematician, a symmetry transformation of a figure is a transformation that leaves the figure *invariant*; i.e., the figure looks the same after the transformation, in terms of position, shape, and orientation, as it did before, even though individual points may have moved.

Because translations are included among the possible symmetry transformations, repeating wallpaper patterns are symmetries. In fact, the mathematics of symmetry is what lies behind the proof of the surprising fact that there are exactly seventeen possible ways to repeat a particular local pattern. The permissible transformations—the symmetries for wallpaper patterns—have to work on the entire wall, not just one part of it. This is what limits the number of symmetries to seventeen.

The proof of the wallpaper-pattern theorem involves a close examination of the ways transformations can be combined to give new transformations, such as performing a reflection followed by a counterclockwise rotation through 90°.

It turns out that there is an arithmetic of combining symmetry transformations, just as there is an arithmetic (the familiar one) for combining numbers. In ordinary arithmetic, you can add two numbers to give a new number, and you can multiply two numbers to give a new number. In the arithmetic of symmetry transformations you combine two symmetry transformations to give a new symmetry transformation by performing one of the transformations followed by the other. The collection of all the symmetry transformations of an object, together with the arithmetic of combining them in this way, is what mathematicians call a symmetry group.

For example, the symmetry group of a circle consists of all rotations about the center (through any angle, in either direction), all reflections in any diameter, and any combination of such. Invariance of the circle under rotations about the center is referred to as rotational symmetry; invariance with respect to reflection in a diameter is called reflectional symmetry.

The arithmetic of symmetry groups works in some ways like the arithmetic of numbers. But there are interesting differences. The discovery of this strange new arithmetic in the latter part of the eighteenth century opened the door to a host of dazzling new mathematical results that affected not only mathematics but physics, chemistry, crystallography, medicine, engineering, communications, and computer technology.

Since it is such a simple example, I'll use the symmetry group of a circle to show how you do arithmetic with a group.

If S and T are two transformations in the circle's symmetry group, then the result of applying first S and then T is also a member of the symmetry group. (Because both S and T leave the circle invariant, so too does the combined application of both transformations.) Mathematicians denote this double transformation by $T \circ S$. The arithmetic of this operation is similar to the arithmetic of addition of numbers in the following three ways.

First, the operation is what is called associative: if S, T, W are transformations in the symmetry group, then

$$(S \circ T) \circ W = S \circ (T \circ W)$$

Second, there is an identity transformation that leaves unchanged any transformation it is combined with. It is the null rotation, the rotation through angle 0. The null rotation, call it I, can be applied along with any other transformation T to yield

$$T \circ I = I \circ T = T$$

The rotation I plays the same role here as the number 0 does in addition ($x + 0 = 0 + x = x$, for any whole number x).

Third, every transformation has an inverse: If T is any transformation, there is another transformation S such that the two combined together give the identity:

$$T \circ S = S \circ T = I$$

The inverse of a rotation is a rotation through the same angle in the opposite direction. The inverse of any reflection is that very same reflection. To obtain the inverse for any finite combination of rotations and reflections, you take the combination of backward rotations and re-reflections that exactly undoes its effect: Start with the last one, undo it, then undo the previous one, then its predecessor, and so on.

The existence of inverses is another property familiar to us in addition of whole numbers: For every whole number m there is a whole number n such that $m + n = n + m = 0$. The inverse of m is just negative m. That is, $n = -m$.

Although we were thinking about symmetries of a circle, everything we just observed is true for the group of symmetry transformations of any figure or object. (This is easily checked.)

In general, whenever mathematicians have some set G of entities (which could be the set of all symmetry transformations of some figure, but need not be), and an operation $*$ that combines any two elements x and y in the set G to give a further element $x * y$ in G, they call this collection a group if the following three conditions are met:

G1. For all x, y, z in G, $(x * y) * z = x * (y * z)$.
G2. There is an element e in G such that $x * e = e * x = x$ for all x in G.
G3. For each element x in G there is an element y in G such that $x * y = y * x = e$, where e is as in condition G2.

These three conditions (generally called the axioms for a group) are just the properties of associativity, identity, and inverses we already observed for combining symmetry transformations of any figure. Thus the collection of all symmetry transformations of a figure is a group: G is the collection of all symmetry transformations of the figure, and $*$ is the operation of combining two symmetry transformations.

It should also be clear that if G is the set of whole numbers and the operation $*$ is addition, then the resulting structure is a group. Alternatively, if G is the set of all rational numbers (i.e., whole numbers and fractions) apart from zero, and $*$ is multiplication, then the result is a group. All you have to do is show that the conditions G1, G2, and G3 above are all valid for the rational numbers when the symbol $*$ denotes multiplication. In that example, the identity element e in axiom G2 is the number 1.

The finite, modular arithmetic discussed in Chapter 6 provides (for any given modulus) another example of a group.

In the branch of mathematics known as group theory, mathematicians see what other properties of groups follow automatically from the three group axioms. Anything that they can show to be a logical consequence of the axioms will be automatically true for any particular group.

For example, condition G2 asserts the existence of an identity element. In the case of addition of whole numbers, there is only one identity: the number 0. Is this true of all groups, or is it something special about whole number arithmetic?

In fact, any group has exactly one identity element. If e and i are both identity elements, applying the G2 property twice in succession gives the equation

$$e = e * i = i$$

Thus e and i must be one and the same.

This last observation means in particular that there is only one element e that can figure in condition G3. Using that fact, we can go on to show that for any given element x in G, there is only one element y in G that satisfies the condition in G3.

Suppose y and z are both related to x as in G3. That is, suppose that

$$x * y = y * x = e \tag{1}$$

$$x * z = z * x = e \tag{2}$$

Then

$$
\begin{aligned}
y &= y * e && \text{(by the property of } e) \\
&= y * (x * z) && \text{(by equation (2))} \\
&= (y * x) * z && \text{(by G1)} \\
&= e * z && \text{(by equation (1))} \\
&= z && \text{(by the property of } e)
\end{aligned}
$$

So y and z are one and the same.

Since there is precisely one y in G related to a given x as in G3, that y may be given a name: It is called the (group) inverse of x and is often denoted by x^{-1}.

One final remark concerning the group axioms is in order.

Anyone familiar with the commutative laws of arithmetic might well ask why we did not include it as a fourth axiom:

G4. For all x, y in G, $x * y = y * x$

The absence of this law meant that in both G2 and G3, the combinations had to be written two ways. For instance, both $x * e$ and $e * x$ appear in G2.

The reason mathematicians do not include an axiom G4 is that it would exclude many of the examples of groups that mathematicians wish to consider. By writing G2 and G3 the way they do, and leaving the commutative law out, the group concept has much wider application than it otherwise would. Groups that satisfy the commutative law are called commutative groups, or sometimes abelian groups after the Norwegian mathematician Niels Henrik Abel.

THREE

When Computers Fail

The P vs. NP Problem

Of all the Millennium Problems, the P versus NP puzzle is the one most likely to be solved by an "unknown amateur"—someone largely untrained in mathematics, possibly someone very young, who is unknown to the mathematical community. All the other Millennium Problems are buried deep within a mass of heavy-duty mathematics, which has to be mastered before you can begin working on the problem itself. This is not the case for the P versus NP problem, which deals with how efficiently computers can perform certain kinds of tasks. Not only is it relatively easy to understand what the problem says, it is possible that all it will take to solve it is one good new idea.[1] And you don't need lots of knowledge to have a good idea, just imagination.

---

1. Note, however, that "possible" does not imply "probable". Despite the simple formulation of the P versus NP problem, I believe that, as with all the other Millennium Problems, when it is finally resolved, the solution will be found by a professional mathematician, using techniques not generally known to the enthusiastic amateur.

Mathematics and Computers:
Rediscovering the Abandoned Child

To most people these days, the computer is a communication device, used to send and receive messages and to obtain information on the Web. But that's not how computers began. As the name suggests, computers were originally developed to do arithmetic—to compute numbers. In fact, they grew out of a theoretical investigation into the mathematical concept of "computability" that predated the technology by fifteen years or more.

In the 1930s, several mathematicians, working largely independently, started to investigate the concept of "computability"—what exactly *is* a computation and which functions from natural numbers to natural numbers can be computed? Not only was this investigation carried out long before modern computing technology became available, but in fact the initial mathematical interest was not motivated by any thought of actually carrying out computations at all, either by machine or by hand. Rather, the question was investigated purely because of its intrinsic mathematical interest.

That interest in computability was aroused by a major discovery made by the Austrian mathematician Kurt Gödel in 1931. Gödel in turn made his discovery in response to a challenge laid down thirty years earlier by David Hilbert, the German mathematician whose 1900 list of major unsolved problems of mathematics inspired the Millennium Problems.

Hilbert was greatly impressed by the success of the axiomatic method for doing mathematics that had become dominant during the nineteenth century. The axiomatic approach to mathematics says that you begin any branch of mathematics by formulating a set of basic assumptions—"axioms"—and then generate all the facts in that branch of mathematics by deducing them logically from the axioms. Thus "truth" reduces to "provable from the axioms." This view of mathematics was first put forward by the ancient Greek mathematician Thales around 700 B.C., and it formed the basis of ancient Greek mathematics. For instance, in his book *Elements*, written around 350 B.C., Euclid developed geometry by first writing down a list of five basic axioms, and then deducing all the theorems (the facts of geometry) from those axioms.

Of course, the success of this method depends on how good a job you make of formulating the axioms. To qualify as an acceptable axiom, a mathematical statement should ideally be relatively simple and sufficiently basic to be "obviously true." This is not always easy to achieve. Euclid's axioms for geometry led to hundreds of years of debate concerning the truth of one of his axioms, the so-called Parallel Postulate.[2] Many critics said that this axiom was too complicated to be taken for an axiom, and over the centuries many attempts were made to try to deduce it from more basic assumptions. The story of how that particular saga led to the development of various "non-Euclidean geometries" (having alternative axioms to the Parallel Postulate) has been recounted on many occasions, so I won't pursue it here, although it is of interest to know that non-Euclidean geometry is exactly what Einstein needed for his theories described in the previous chapter. Rather, more to our present theme is the observation that when you are trying to write down axioms, it can be difficult to identify all of the basic, "self-evident" assumptions that should be included in your list. Euclid missed several subtle but crucial assumptions you need to do geometry, assumptions that he used throughout *Elements*, and it was not until Hilbert took a closer look in the late nineteenth century that a complete set of axioms was written down.

Following his success in formulating an adequate set of axioms for Euclidean geometry, Hilbert proposed that the same could—and should—be done for any other branch of mathematics. The search for axioms for the various branches of mathematics became known as the Hilbert Program.

Behind the Hilbert Program was an unstated assumption

---

2. The Parallel Postulate says that, given a straight line and a point not on that line, you can draw exactly one straight line through that point that is parallel to the given line.

that the method worked. That is, in any area of mathematics, it is theoretically possible to write down a set of basic assumptions—axioms—from which all of the facts in that branch of mathematics can (in principle) be deduced. In 1931, Kurt Gödel rocked the mathematical world with his discovery that this assumption was false. He proved that in any part of mathematics that includes elementary arithmetic (which means practically any remotely useful part of mathematics), no matter how many axioms you write down, there will *always* be some true statements that cannot be proved from those axioms. This result, which completely devastated the Hilbert Program, is known as the Gödel Incompleteness Theorem. In everyday language, it says that no matter how hard you try, your set of axioms will always by incomplete—they won't be sufficient to prove all the true facts. In mathematics as in life, parts of the truth are destined to remain forever elusive.

Gödel proved this result by showing how to translate questions about provability to equivalent questions about computability of certain functions from natural numbers to natural numbers. (This is why his theorem worked only for parts of mathematics that include some arithmetic. The axioms had to make it possible to do that arithmetic.) He showed that in any axiomatized system, there will always be some functions that cannot be computed within that system. To do this, he had to develop a formal theory of the concept of a "computable function."

Building on Gödel's work, other mathematicians started to investigate the concept of computability, to try to see just which functions can be computed and which cannot. (Let me repeat that no one was concerned about carrying out any computations. No actual numbers were involved. This was a purely theoretical study about what computations could *in principle* be carried out.)

With hindsight, it is fascinating to see theorems proved by the mathematicians Stephen Kleene, Alan Turing, and others, that established—long before such things—the theoretical possibility of building *programmable* computers, machines that could be programmed to perform different computations. The theoretical ideas developed in the 1930s and early 1940s played a major role in the early development of computers in the 1940s and 1950s, with some of the mathematicians who worked on the theory (in particular Turing and John von Neumann) playing a major role in developing the new technology.

And then a strange thing happened. Having developed the mathematical theory that led to the computer, and having helped build and program the very first computers, the mathematicians largely lost interest in their brainchild. It's not hard to see why this happened. While the development of computer technology—both hardware and software—required some mathematical ability, and often used mathematical notation, most of the work was not really mathematics as such. Hence most mathematicians were simply not interested. And when it came to *using* computers, the vast majority of mathematicians worked (as they do today) on problems that required no heavy-duty numerical calculation, so the computer did not affect them the way it did, say, physicists or chemists. (The situation changed quite a bit in the late 1980s, when sophisticated computer systems were developed for doing algebra, calculus, and other branches of symbolic mathematics.)

Still, from the very start there were *some* mathematicians who were interested in finding ways to use computers to help solve mathematical problems, and a number of new branches of mathematics sprang into being because of computer technology—including numerical analysis, approximation theory, computational number theory, and dynamical systems theory. There were also some mathematicians who studied the concept of computation with a view to improving the way real computers were used. Some of those early studies led to the emergence of the new discipline of computer science—with mathematical subdisciplines such as formal language theory, theory of algorithms, database theory, artificial intelligence, and computational complexity. It is in this last subject that we find the fourth Millennium Problem. The person who did most to establish this question as the preeminent problem of theoretical computer science was a young American named Stephen Cook.

The Cook's Tale

Stephen Cook was born in Buffalo, New York, in 1939. When he was ten, his family moved into the country, taking up residence in a farmhouse near to the small town of Clarence, New York. Clarence was the hometown of Wilson Greatbatch, the inventor of the implantable heart pacemaker, and the proximity of this local celebrity inspired Stephen to become an electrical engineer. After whetting his appetite by working in Greatbatch's workshop—the converted attic of a barn—during the summers, soldering the newly invented transistors into circuits, in 1957 Stephen went to the University of Michigan to study electrical engineering.

During his first year, Cook took a one-credit course in computer programming and was hooked. Together with a friend, he wrote a program to test Goldbach's Conjecture that every even number greater than 3 is the sum of two primes. (See page 29.) Electrical engineering was abandoned. Stephen declared a major in mathematics—back in the 1950s, computer science did not yet exist as a separate discipline, although a few mathematics departments offered courses related to it. Cook took all the computing-related courses that Michigan offered. He was particularly intrigued by Turing's solution to the Halting Problem, which says that there is no program that can examine any given program and tell whether that program will stop in a finite amount of time.

When he graduated from Michigan, in 1961, Cook went to Harvard to do a Ph.D. in mathematics. His intention was to study algebra, but he soon found himself influenced by the logician Hao Wang, then teaching in Harvard's applied science division. Wang was working in the new area of automatic theorem proving, a subfield of the equally new subject that John McCarthy had ambitiously named artificial intelligence.

While at Harvard, Cook also encountered Michael Rabin's groundbreaking work in complexity theory. Complexity theory analyzes computational processes to see how efficiently they can be carried out. We'll learn more about how this is done in just a moment.

After completing his doctoral thesis in 1966, Cook took a position at the University of California at Berkeley, remaining there for four years before moving to the University of Toronto (his present home) in 1970. A year later, he published his paper *The Complexity of Theorem Proving Procedures*, in which he introduced a new theoretical concept he had discovered, called NP completeness. The rest, as they say, is history. As a result of his discovery, Cook was subsequently elected a fellow of the Royal Society of Canada and a member of the US National Academy of Sciences.

The concept of NP completeness provided complexity theorists with a powerful tool to analyze computational tasks. Although in his 1971 paper Cook proved only that a single highly artificial and obscure problem of propositional logic was NP complete—and hence almost certainly could not be solved effectively on a computer—within a few months Richard Karp of the University of California at Berkeley had shown that twenty-one other problems are also NP complete, including some highly practical problems of considerable interest to industry. Since then, the list of NP complete problems has been extended into several hundreds, possibly thousands, among them almost all of the computational tasks that industry cares most about. All of which came as something of a surprise to Cook himself, who, many years later, said: "I thought NP-completeness was an interesting idea—I didn't quite realize its potential impact."[3]

So what exactly is this idea of NP completeness that Cook dreamt up in 1971? The best way to explain it is with a simple example.

---

3. D. Shasha and C. Lazere, *Out of Their Minds: The Lives and Discoveries of 15 Great Computer Scientists*, New York: Copernicus, 1998, p. 148.

The Ambitious Traveling Salesman

Imagine you are a salesman, based in—where else?—Springfield. You have to visit three cities by car, Oldtown, Midtown, and Newtown, starting and finishing in Springfield. To save gas and time, it makes sense to plan your route so that the total distance traveled is as small as possible. So you get out your travel planner and look up the shortest road distances between each pair of cities. Here is what you find, laid out in a table:

	Springfield	Oldtown	Midtown	Newtown
Springfield	0	54	17	79
Oldtown	54	0	49	104
Midtown	17	49	0	91
Newtown	79	109	91	0

(The distance from Newtown to Oldtown is 5 miles further than the distance from Oldtown to Newtown because of a system of one-way streets in Newtown. For all the other pairs of cities, the distance is the same in both directions.)

Now all you have to do is order the three cities to minimize the total distance traveled. How easy is this? Well, for each ordering of the cities, it's a simple matter to compute the total distance. All you have to do is read off three numbers from the table and add them together. If you do this for each possible route—for each possible ordering of the cities—then by looking through the answers, you can find one that gives the least total. When you do this, here is what you find:

Route	Total mileage
S-O-M-N-S	54 + 49 + 91 + 79 = 273
S-O-N-M-S	54 + 104 + 91 + 17 = 266
S-M-N-O-S	17 + 91 + 109 + 54 = 271
S-M-O-N-S	17 + 49 + 104 + 79 = 249
S-N-O-M-S	79 + 109 + 49 + 17 = 254
S-N-M-O-S	79 + 91 + 49 + 54 = 273

Clearly, the optimal route is S-M-O-N-S, with a total mileage of 249 miles.

Like most simple examples, this is not very realistic. With only three cities to visit, there's not a lot of difference between any of the routes, and it hardly seems worthwhile doing the calculation. But for a more ambitious salesman with more cities to visit, a number of small differences could add up to quite a lot.

Suppose you were in fact faced with visiting ten cities. In this case, you'd probably decide to use your computer to do the math, setting out the mileage table as a spreadsheet and writing a small macro to enumerate the different routes and work out the totals. In fact, you'd better do it that way, or else be prepared to spend a long time at your desk. With ten cities to visit, the number of different routes is 3,628,800.

To see where this number comes from, look back at the first example, with three cities to visit. There are 3 possibilities for the first city. From there, there are 2 cities left to visit next. And then there is just 1 final city to visit before you head back home to Springfield. So, the total number of different routes is

$$3 \times 2 \times 1 = 6$$

These are the six different routes enumerated on the previous page.

For 10 cities, there are 10 possibilities for the first city, then 9 cities left to go to next, then 8 that you might visit third, and so on, so the total number of different routes is

$$10 \times 9 \times 8 \times 7 \times 6 \times 5 \times 4 \times 3 \times 2 \times 1 = 3,628,800$$

The expression on the left of the equal sign here comes up a lot when you have to see how many different ways there are to do various things, so mathematicians have a special notation for it. They write it as

$$10!$$

You don't read this aloud as "ten" in an excited or startled voice, nor do you say "ten exclamation mark." You read it as "ten factorial."

Thus, the expression "10!" (10 factorial) is an abbreviation for the product

$$10 \times 9 \times 8 \times 7 \times 6 \times 5 \times 4 \times 3 \times 2 \times 1$$

That is,

$$10! = 3,628,800$$

In words, "ten factorial equals three million six hundred and twenty-eight thousand, eight hundred."

That's a lot of routes to look at for a tour of a mere ten cities. In fact, suppose you worked out the total mileage for each one of these routes, and that it took you exactly one minute to work out one total. If you worked for eight hours a day, with no breaks, five days a week, for fifty-two weeks of the year, it would take you just over twenty years to complete the task! That last exclamation mark is just that: It's there to express surprise. Twenty years for just ten cities.

Add one more city and the number of possible routes jumps to almost 40 million:

$$11! = 11 \times 10 \times 9 \times 8 \times 7 \times 6 \times 5 \times 4 \times 3 \times 2 \times 1 = 39,916,800$$

Obviously, you don't try to do this kind of computation by hand, you use a computer. But because the factorial numbers grow large so fast, it doesn't take many cities before even the most powerful computer will be overwhelmed. For instance, 25! is just shy of a daunting 16 followed by 25 zeros:

$$25! = 15,511,210,043,330,985,984,000,000$$

Since a tour of 25 cities would not be at all unrealistic for a professional salesman, it's obvious that the Traveling Salesman Problem, as our task is called, presents us with a serious challenge. If at first it seemed so straightforward, that's because it is straightforward. There's no difficult math involved. All you have to do is add strings of numbers together and compare the totals. What makes this difficult—indeed impossible for more than a handful of cities—is the sheer magnitude of the number of routes. Go from 10 cities to 11, and the number of possible routes goes up elevenfold. Add one further city and the number of routes goes up again by a factor of 12. One more city, to give a tour of 13 cities, and the number of routes leaps up yet again, this time by a factor of 13. It seems so innocent: just one more city. But with each city added, the number of routes jumps up by a factor that itself increases.

Mathematicians say that a numerical pattern that increases in this way, where the rate of growth at any stage is roughly proportional to the number at that stage, exhibits exponential growth. The factorial numbers grow exponentially, and that's what makes the Traveling Salesman Problem such a killer.

Will an Approximate Answer Do?

Of course, trying to find the shortest route by listing all possibilities and comparing them is a very naive way of trying to solve the problem. Maybe there is a better way? Because the Traveling Salesman Problem is important in industry and business, mathematicians have expended enormous energy trying to find other methods. Broadly speaking, their approaches have fallen into two categories. Both involve heavy-duty mathematics.

One strategy is to settle for an approximate answer. Instead of looking for the route that has the least total mileage, you look for one that comes within, say, 5% of the optimal route. With this approach, there are methods that work for most real-world situations.

The other strategy is to go for an exact answer, but to look at the overall geography of the cities and try to make use of special features of the layout to cut down the number of possible tours that have to be examined. For example, in general you should avoid itineraries that have you travel directly from the city farthest east to the city farthest west. An approach that takes account of the geography can make sense when you are faced with a particular set of cities for which it is worth expending considerable effort trying to find a solution. The obvious disadvantage is that the answer you get applies just to that particular set of destinations. Add or remove one city, and you have to start all over again. Moreover, it often takes enormous ingenuity to make this approach work.

One of the best results so far was obtained in 1998, when a team of mathematicians found the shortest route that visits all 13,509 cities in the United States having a population of 500 or more. It took three and a half months of nonstop computation on a network of three high-performance, scientific multiprocessor computers supported by 32 Pentium PCs. (This particular problem seems to be of no real importance. The researchers chose it to see just what could be done when the geography is taken into account, such as eliminating obviously poor routes at the outset.)

Leaving aside approximate solutions and answers for some specific collections of cities, however, the fact is that no one knows a practical solution for the Traveling Salesman Problem. None of the known methods is significantly better than listing all the possible tours and comparing the totals, which, as we have seen, is hopelessly ineffective for all but the smallest of tours.

Enter the Theoreticians

The Traveling Salesman Problem was first introduced in the 1930s by the Viennese mathematician Karl Menger. It was not long before mathematicians discovered other, equally important problems in industry and management that similarly resisted solution.

For instance, there is the Process Scheduling Problem. Here you are give a number of tasks that have to be performed, say in a manufacturing plant. Some of the tasks depend on the completion of others, while some tasks are independent. Can you divide them up into groups so that the total time taken to complete all the tasks is as short as possible? Each group of tasks will be tackled as a process, one task after another, but the individual processes can be performed at the same time. (Modern auto-mobile manufacture is an excellent example. The engine can be assembled at the same time as the body, but many of the individual tasks involved in assembling the engine have to be performed in a particular order.)

As with the Traveling Salesman Problem, once you have split the tasks into groups, each of which can be carried out at the same time as all the other groups, the math involved in calculating the total time is trivial. All you have to do is add together the times taken to perform the individual tasks in each group (to determine the time it takes to complete that group) and then identify the maximum of those times. That longest time is how long it takes to complete all the tasks using that particular grouping. The problem is that in order to find the schedule that takes the least time, you have to do this for all possible groupings, and as with the number of possible tours in the Traveling Salesman Problem, this grows exponentially as the number of tasks increases.

With the industrial mathematicians struggling largely in vain to solve these problems, it was not long before the theoreticians—the pure mathematicians—took notice. As is so often the case, when the pure mathematicians began to think about the problem, they asked a very different question. Suppose, they said, there simply is no effective way to solve the Traveling Salesman Problem (except in an approximate way). The question then is, can you prove *this*? If you can, then you know at least that it is pointless devoting large amounts of time, brainpower, and computational resources to trying to solve it.

It's not enough to say that a lot of bright people have tried hard for a number of years and failed. Maybe they simply haven't yet come up with the right idea. To prove, conclusively, that it's a waste of time and effort trying to solve the problem, you'd have to come up with a solid proof that there is no way to solve it that's significantly better than the brute-force method of trying all the possibilities.

This was a definite shift in focus, from trying to solve a particular problem on a computer to looking at ways of solving problems with computers. The Traveling Salesman Problem is a computational task of working out routes. The problem the theoreticians started to look at was about how *efficiently* a particular task can be carried out on a computer.

How Many Steps Does It Take to Solve a Problem?

The first problem the theoreticians faced was finding a way to measure the length of time it would take to perform a particular task on a computer. For example, how long would it take to find the shortest tour of a particular collection of cities? Clearly, the answer depends on (at least) two things: the computer used, in particular its speed and memory capacity, and the number of cities in the collection. Given all the attention that the press regularly gives to the speed and capacity of the latest computer on the market, you may be surprised to learn that, from a theoretical standpoint, this is not at all a significant factor. What matters is the number of cities.

Obviously, the more data a problem has, the longer it will take to perform the calculation. But how much longer? More precisely, if the amount of data is increased by a certain amount, by how much does the computation time increase? For instance, if we double the amount of data, does the computation time double? Does it triple? Does it increase by a factor of 10? Or is the increase even more dramatic?

Since our interest is in the proportional increase in computation time, whether it doubles, triples, etc., it doesn't matter what the actual computation times are. In which case, all we need to do is see how many basic steps are involved in the computation. This changes the problem from measuring time to counting basic computational steps. (In essence, this is why this analysis does not depend on the type of computer used.)

What is a basic step? If we were analyzing the way people do arithmetic, a basic step would be the addition or multiplication of two single digits. These are the basic number facts that we all have to learn as children. Once we have mastered those facts, we can add, multiply, subtract, or divide any pair of numbers by following a standard procedure. Allowing for carries, using that procedure to add together two N-digit numbers (where N can be 2, 3, 4, etc.) involves at most $3N$ basic steps. For example, adding together two 4-digit numbers requires $3 \times 4 = 12$ basic steps.

Using the standard method for multiplying two N-digit numbers involves N^2 basic integer-pair multiplications (using the multiplication tables we all learn at school) plus at most N additions to take care of carries. Altogether, that's at most $N^2 + N$ basic operations. Since the object is to get a sense of how many steps a particular computation might take, mathematicians simplify the algebra by observing that the value of the expression $N^2 + N$ is always less than $N^2 + N^2$, which is equal to $2N^2$. Thus, multiplication of two N-digit numbers involves fewer than $2N^2$ basic operations. For example, multiplying two 4-digit numbers requires fewer than $2 \times 4^2 = 32$ basic steps.

We can analyze computer arithmetic in roughly the same way. Since all computers work in essentially the same way at their most basic level, one approach is to take a single binary step of the CPU (the central processing unit, the part of the computer that does the actual computation). But there is no need to go down to such a level. The analysis works out essentially the same if you take as a basic operation the addition or multiplication of a pair of decimal numbers within the computer's normal range.

Whether we look at mental arithmetic analyzed in terms of single-digit operations or computer arithmetic analyzed in terms of basic arithmetic operations, addition provides an example of what is called a "linear-time process." A linear time process is one that for data of size N (e.g., in the case of mental addition, the two numbers to be added each have N digits) takes at most $C \times N$ basic steps to complete, where C is some fixed number ($C = 3$ in the case of mental addition).

Since we just agreed to carry out the analysis in terms of basic steps rather than time of computation, we should perhaps call this "linear basic steps" rather than "linear time." But since the original aim of analyses of this kind was to see how long it would take a computer to perform a particular task, the name "linear time" was used first, and it has stuck. The same convention is used in talking about other computational tasks. Since any basic operation will (we can assume) take the same fixed amount of time, the number of basic steps corresponds directly to the time the computation takes, so the historically motivated terminology does not conflict with the mathematics.

The word "linear" in the phrase "linear time" indicates that if you draw a graph of the number of steps against the size of the data, it will be a straight line. (Its equation will be $S = CN$, where S is the number of steps.)

In contrast, multiplication is a "quadratic-time" process. In general, a process is said to run in quadratic time if for data of size N, it takes at most $C \times N^2$ steps to complete, where C is some fixed number ($C = 2$ in the case of mental multiplication).

A more general concept than linear- or quadratic-time processes is that of a "polynomial-time process." A polynomial-time process is one that for data of size N takes no more than $C \times N^k$ basic operations, for some fixed numbers C and k.

All of the four arithmetic operations—addition, subtraction, multiplication, and division—are polynomial-time processes.

When presented with a computational process, the theoreticians look for an algebraic expression (for example, CN or CN^2 or CN^k) that gives an upper estimate of the number of basic steps the process requires for data of a given size N. They call such an expression the "time complexity function" for that process. The polynomial-time processes are those having a time complexity function that is a polynomial expression, i.e., an algebraic expression such as CN or CN^2 or CN^k.

Longer Than the Age of the Universe: Polynomial Time Versus Exponential Time

Roughly speaking, the polynomial-time processes are the ones that a computer can deal with effectively. I say "roughly speaking" because if the fixed numbers C and k are very large, for instance if k is many thousand, then the process can involve so many steps that the computation could take up the entire lifetime of the universe. In practice, however, the polynomial-time processes that tend to arise in everyday life have fairly modest values of C and k—indeed k is generally a single-digit number—and so they really can be handled effectively on a computer.

The real value of the "polynomial-time" category comes from contrasting it with computational tasks that require "exponential time." These are processes that, when presented with data of size N, require 2^N or more basic steps to complete. (Strictly speaking, the definition the experts use is broader than this, but that's a technical point we can safely ignore here.) For example, the simple search procedure to solve the Traveling Salesman Problem requires at least $N!$ basic steps to find the answer, which is considerably more than 2^N steps.

What makes exponential-time processes all but impossible to run on even the most powerful computer is the rate at which the number 2^N grows as N increases. To get some idea of this growth, imagine an ordinary chessboard. Suppose we number the squares on this chessboard sequentially, starting with 1 in the top left-hand corner and proceeding row by row down to number 64 in the bottom right-hand corner. Now imagine that we start putting dollar coins on the squares of the chessboard. On square number 1 we put 2 dollar coins (2^1), on square 2 we put 4 (or 2^2), on square 3 we put 8 (or 2^3), and so on, on each successive square putting exactly twice as many coins as on the previous one. On the last square we will place exactly 2^{64} dollar coins. How high do you think this stack will be? Six feet? Fifty feet? More? In fact, it will be about 37 million million kilometers high. It would stretch way beyond the moon (a mere 400,000 kilometers away) and the Sun (150 million kilometers from Earth), and would reach the nearest star, Proxima Centauri.

For the first few squares on the chessboard, the numbers of coins in each pile don't seem too large. It's when N starts to get larger that the piles begin to leap up dramatically. Likewise, for a sufficiently small quantity of data, it might be possible to run an exponential-time process and get an answer. For instance, we solved the Traveling Salesman Problem by hand for a tour of three cities. But as the amount of data grows larger, eventually there is simply not enough time to complete the computation. For practically all exponential-time processes that arise in industry and business, handling even quite moderate-sized—and highly realistic—data would take the world's fastest computer longer than the entire age of the universe. The following table gives comparisons of the times it would take a moderately fast computer—one that can carry out one million basic arithmetical steps a second—to run processes having various time-complexity functions.

Time-complexity function	Size of data: N				
	10	20	30	40	50
N	0.00001 sec	0.00002 sec	0.00003 sec	0.00004 sec	0.00005 sec
N^2	0.0001 sec	0.0004 sec	0.0009 sec	0.0016 sec	0.0036 sec
N^3	0.001 sec	0.008 sec	0.027 sec	0.064 sec	0.125 sec
2^N	0.001 sec	1.0 sec	17.19 min	12.7 days	35.7 years
3^N	0.059 sec	58 min	6.5 years	3,855 centuries	200,000,000 centuries

For the first three rows the processes run in polynomial time. The computation time goes up as the data size increases, but the increases are steady, and even for data of size 50 the processes take only a fraction of a second.

The last two rows show processes that run in exponential time. Here the computation time increases dramatically. For processes that run in time 2^N, once the data size gets up to about 40, the computation runs for days, and the data size has to go up to only 50 for the process to require over 35 years to find the answer. For processes that run in time 3^N, data of size 40 requires almost 4,000 centuries to handle, and data of size 50 needs a whopping 200 million centuries.

The table shows the enormous gulf that separates polynomial-time processes from exponential-time processes. Clearly, if the only way you know how to solve a particular problem is by using an exponential-time process, then you're not going to be able to solve that problem for all but the smallest quantities of data. In particular, for all but relatively small values of N, $N!$ is much bigger than 3^N, so the examine-all-possibilities approach to the Traveling Salesman Problem is a nonstarter.

A Finer Mesh

The huge gulf between polynomial-time and exponential-time processes illustrates an obvious weakness of this classification: It is far too crude. Realizing this, mathematicians looked for intermediate measures of process complexity. They observed that for processes such as the simple search method for solving the Traveling Salesman Problem or the Process Scheduling Problem, the difficulty was not a result of a complicated computation. On the contrary, the computation was extremely simple. What made the problem all but impossible to solve was the sheer number of possibilities that had to be examined, for which the same extremely simple computation had to be repeated over and over. Because a person or a (digital) computer had to carry out those calculations serially—i.e., one after another—the entire process took a prohibitively long time to complete.

To try to distinguish such processes from those that involved a genuinely complicated calculation, mathematicians came up with a third classification: nondeterministic polynomial-time processes, or NP processes for short. Since ordinary computers are deterministic—they work in a completely predictable fashion, following rules stipulated in advance—the use of the word "nondeterministic" should give a hint that the new concept is a theoretical one having little to do with actual computing. Here is the general idea.

Imagine you have a computer that at certain stages in a computation can make a completely random choice between a number of alternatives. For example, presented with an instance of the Traveling Salesman Problem, the computer can randomly pick one of all the possible tours that the salesman could follow. To solve the problem, the computer picks one tour and calculates the corresponding total distance. If this were done in practice, the probability is overwhelming that the resulting tour would not be the shortest. But suppose that this particular computer is blessed with incredible luck, so that it always makes the best choice. Then it would solve the problem in polynomial time. The capability of making a lucky random guess circumvents the problem of the prohibitively large number of possibilities.

In general, we say that a problem or task is of type NP if it can be solved or completed in polynomial time by a nondeterministic computer that is able to make a random choice between a range of alternatives and moreover does so with perfect luck. (But note that the computer has to check that its guess is correct. The essence of the class NP is that it is *only* the sheer number of possibilities that is the problem. For an NP problem, checking a given solution to see whether it is correct has to be doable in polynomial time.)

Intuitively, the NP problems are intermediate between the polynomial-time problems (P problems for short) and the exponential-time problems. Because it is built on the totally unrealistic idea of a computer that can invariably make the best random choices, the NP concept is purely theoretical. It nevertheless turns out to be of considerable importance. One reason (I'll give another shortly) is that most of the exponential-time problems that arise in industry and management are of type NP. What makes them hard to solve is not that the computation is complex, but that a relatively straightforward computation has to be carried out for an overwhelmingly large number of virtually identical cases.

When the NP classification was first introduced in the 1960s, computer scientists assumed that the classes P and NP are not identical—that although every P problem is automatically an NP problem, there are some NP problems that are definitely not in P. The reason is that there seems to be no way a standard computer running a polynomial-time algorithm could perform as well as an imaginary nondeterministic computer can get by making perfect guesses. For example, the experts thought it likely that the Traveling Salesman Problem simply cannot be solved in polynomial time without the perfect guessing ability of a hypothetical nondeterministic computer.

It was only a matter of time, everyone supposed, before someone would prove that P and NP are distinct classes of problems, by exhibiting a particular NP problem that was provably not of type P—if not the Traveling Salesman Problem, then some other task. But that did not happen. Nor was anyone able to prove the opposite result, that P and NP are in fact the same. Thus the P versus NP problem was born.

By this stage—the late 1960s—there was a considerable amount at stake. Many important problems in industry and management were known to be NP. A proof that P is the same as NP would undoubtedly have unleashed a massive effort to find efficient procedures for solving those important problems. (I should point out that proving NP is the same as P might not, in itself, lead to efficient procedures for solving specific NP problems. All it would show is that any NP problem could, in principle, be solved by a polynomial-time procedure. It would not necessarily provide any hint as to what such a procedure might look like.)

This is where Stephen Cook entered the picture. With his famous 1971 paper, Cook upped the ante considerably on the P versus NP problem, and in so doing provided a second reason for the importance of the NP classification.

Cook demonstrated that there was one particular NP-problem with a curious property. If this particular problem could be solved by a polynomial-time procedure, then so too could any other NP problem! The exact nature of Cook's problem need not concern us here. In general terms, it is a problem about what kinds of tasks can be performed on a nondeterministic computer. Cook proved his result by showing how any NP problem could be translated to his particular problem, so that if his problem could be solved in polynomial time, so, via the translation, could the given one. The name Cook gave to this strange property was NP-completeness. According to Cook, an NP problem was said to be NP-complete if the discovery of a polynomial-time procedure to solve it would imply that every problem in NP could be solved by a polynomial-time procedure. Although Cook's problem was a highly theoretical one from formal logic, it was not long before Richard Karp and others were able to show that many other more familiar NP problems also had this property of NP-completeness, including the Traveling Salesman Problem and the Process Scheduling Problem.

The introduction of the notion of NP completeness, and the discovery that most of the important NP problems are NP complete, was arguably a major blow to industry, for whom the discovery of efficient procedures for solving problems like the Traveling Salesman Problem could have led to billions of dollars of increased profits. The issue was not that NP completeness meant that a problem *definitely* could not be solved efficiently. After all, with the P versus NP problem still unsettled, it could still turn out that P and NP are the same, in which case any NP problem could in principle be solved in polynomial time. Rather, the proof that a particular problem is NP complete is a measure of just how hard it is, and just how unlikely it is that you will find a polynomial-time procedure to solve it. Here's why.

Because problems like the Traveling Salesman Problem or the Process Scheduling Problem are so important, over many years, a great many very talented researchers have devoted many hours to trying, and failing, to find efficient ways to solve them. Suppose now you discover that your favorite NP problem is in fact NP complete. What that tells you is that your problem is just as hard as all those other problems that all those other people have been unable to solve. Consequently, most experts take a proof that a particular problem is NP complete as an adequate reason not to devote time and effort to trying to find a full solution. Instead, they devote their energies to looking for good approximate general solutions or one-off solutions to specific instances of the problem. Thus, despite its highly artificial nature, the NP classification does help managers decide where to invest their research efforts.

And yet lurking behind everything is the still unresolved P versus NP problem. A proof that P and NP are the same would, in principle, render all the work on NP completeness a waste of time. (I say "in principle" because a proof that P and NP are the same would still leave you with the task of finding actual polynomial-time procedures that can solve the various NP problems.)

Such a proof would also have major consequences for Internet security. As we saw in Chapter 1, in the mid 1970s, mathematicians and computer scientists developed a powerful new method for encrypting electronic messages sent over an open computer network. The security of the method—called RSA encryption, after the initials of the three mathematicians who developed it—depends on the problem of breaking the code not being in the class P. Thus, although the code could, in principle, be broken, it would take the fastest computers many years to do so. Breaking the code is, however, an NP problem. (As with the Traveling Salesman Problem, the difficulty is the vast number of possibilities involved.) A proof that the code-breaking problem was in fact in P would immediately compromise the method.

The code-breaking problem for RSA encryption is not known to be NP-complete (and probably isn't), so perhaps a polynomial-time solution to that problem could be developed without proving the identity of P and NP. But going the other way, a proof that P equals NP would imply at once that the code-breaking problem for RSA could be solved in polynomial time, and thus would throw the entire Internet security system into doubt. Since we do not, at present, know of any way of ensuring the security of open Internet communications that does not depend on the effective impossibility of solving an NP problem, the current dependence of the Western economies on secure electronic communications over the Internet demonstrates just how high are the P = NP stakes.

Is It True or False?

Is P the same as NP or not? The discovery that a great many problems are NP complete means that mathematicians have many ways to try to prove that P = NP. Find a polynomial-time procedure that solves any NP-complete problem and P = NP follows immediately. For example, a polynomial-time procedure to solve the Traveling Salesman Problem will constitute a proof that P = NP.

Nevertheless, the smart money is probably still on P being different from NP. To prove this, you would have to find an NP problem for which you could prove that there was no polynomial-time procedure to solve it. The problem could be one already known. For example, if you could prove that there is definitely no polynomial-time procedure to solve the Traveling Salesman Problem, you would have shown that P and NP are different.

This is not as easy as you might think. It's not enough to take some particular procedure that solves the Traveling Salesman Problem and show that it is not a polynomial-time procedure. Nor is it enough to prove that none of the procedures that have been developed so far runs in polynomial time. Rather, you have to show that there can be no procedure that solves the problem in polynomial time. That means that your proof has to take account of any procedure that can solve the problem, not just the ones that are known but any procedure that might—or might not—be discovered in the future.

It might seem strange to outsiders, but mathematicians have on several occasions been able to prove results about such unknown collections of objects or procedures. Cook's proof of NP completeness was such a result. He showed that if his particular NP problem could be solved in polynomial time, so too could any other NP problem, including all NP problems not yet discovered. Nevertheless, in the case of proving P ≠ NP, no one has come close to showing that there is some NP problem that no polynomial-time procedure can solve. That's why the P versus NP problem is a Millennium Problem.

Still, as I said at the beginning of this chapter, of all the Millennium Problems, the P versus NP puzzle is the one most likely to be solved by an unknown amateur. Not only is it relatively easy to understand what the problem says, it is possible that all it will take to solve it is one good new idea. Many years ago, I spent about a week thinking about the problem, looking for that one new idea. I didn't get far. In fact, I made no progress at all. The approach I adopted—and I was trying to prove that P and NP are different—was to try to formulate a computational problem that was obviously NP but for which I could prove that no polynomial-time procedure could solve it. The idea—which I have no doubt many other mathematicians have tried—was to design my NP problem so that *by its nature* it could not be solved in polynomial time. This NP problem would not be one of the standard ones coming from industry. I expected it would have a strange, artificial appearance—perhaps not unlike Cook's original NP-complete problem—since the idea was to build into the problem enough information to allow me to prove that it could not be solved in polynomial time.

As I say, I got nowhere with this approach. And I mean nowhere. I could not find a way to build in a property that I could use to show that the problem could not be solved in polynomial time. Nevertheless, I still think that this is how the P versus NP problem will eventually be solved. If you want to give it a try—which is not something I would recommend to a non-professional for any of the other Millennium Problems—then the best I can do is wish you good luck!

FOUR

Making Waves

The Navier–Stokes Equations

V isitors to the Eiffel Tower are usually too busy looking upward to notice that on each of the four faces is a plaque
bearing a list of names. But for those who do, a closer inspection
reveals that, contrary to expectation, they are not individuals associated with the erection of the tower. Rather, when Gustave
Eiffel built his famous monument in 1889, he chose to honor 72
nineteenth-century French scientists by listing their names on the
tower.

Among the 18 names listed on the first facade (the one opposite the Trocadero) are several famous mathematicians, most
notably Lagrange, Laplace, and Legendre. Also on that plaque
you will find the name of Claude Louis Marie Henri Navier. But
unlike those others, Navier's name did not appear because of
his contributions to mathematics or theoretical physics, which is
how he is remembered today. He was better known during his
lifetime as one of France's most famous engineers—a designer
and builder of bridges—and a prominent public figure. Navier
was a close friend of the famous French philosopher Auguste
Comte, one of the founders of sociology, and from 1830 until

his death in 1836, he was employed by the French government as a consultant, advising on how to use science and technology to improve the country.

How did this famous bridge builder and government advisor come to have his name attached to one of the seven Millennium Problems?

Despite his fame as an engineer, Navier was also trained as a mathematician, even studying for a while under the great mathematician Joseph Fourier at the prestigious Ecole Polytechnique in Paris. It was around 1820, while Navier was teaching engineering and applied mathematics at the Ecole des Ponts et Chaussées (College of Bridges and Causeways) that he began to think about the mathematics of flowing fluids. He discovered the now famous Navier–Stokes equations in 1821 and 1822.

Their equations are central to our mathematical description of fluid flow. To understand where they come from, we need to go back two hundred years.

During the first half of the nineteenth century, the Swiss mathematician Daniel Bernoulli showed how to adapt the methods of calculus to analyze the way fluids move when subjected to various forces. Building on Bernoulli's work, his countryman Leonhard Euler formulated a set of equations whose solution describes precisely the motion of a hypothetical viscosity-free fluid.

In 1822, Navier amended Euler's equations to cover the more realistic case of a fluid having some degree of viscosity. Navier's mathematical reasoning was flawed, but by good fortune (or by an engineer's good intuitions) the equations he ended up with were correct. A correct derivation was obtained a few years later by an Irish mathematician named George Gabriel Stokes.

Born in County Sligo in Ireland in 1819, the mathematically precocious young Stokes entered Pembroke College, Cambridge, in 1837, graduated with top marks in mathematics in 1841, and was awarded a scholarship to stay on and carry out research.

From the start, Stokes concentrated on using methods of calculus to try to understand the flow of fluids. He rediscovered (in his case, with correct reasoning) the equations Navier had formulated twenty years earlier. In fact, Stokes took the theory much further than Navier had. He went on to an illustrious career that led to his appointment in 1849 as the Lucasian Professor of Mathematics (the prestigious position held formerly by Isaac Newton and today by the astrophysicist Stephen Hawking) and, in 1852, to his election to the Royal Society. (Although Stokes was not commemorated by having his name on a world famous monument the way Navier was, he *has* achieved what many would regard as an even greater form of memorial: Craters on both the Moon and on Mars have been named after him!)

With the work of Navier and Stokes, by the end of the nineteenth century it looked as though mathematicians were on the brink of working out a complete theory of fluid flow. Such a theory could reasonably be expected to have many applications. For example, understanding how fluids flow over surfaces might result in improvements in the design of ships and airplanes. Perhaps it could help us to understand the way the heart works and the way blood flows through our arteries and veins—maybe leading to the development of life-saving medical equipment.

There is just one problem. No one has been able to find a formula that solves the Navier–Stokes equations. In fact, no one has been able to show in principle whether a solution even exists! (More precisely, we don't know whether there is a *mathematical* solution—a *formula* that satisfies the equations. Nature "solves" the equations every time a real fluid flows, of course.) The most significant lesson we have learned to date is that the mathematics of fluid flow seems to be extremely hard.

And yet for two hundred years the mathematical path that led to the formulation of the equations seemed reasonably straightforward, with little indication that progress would soon come to a virtual standstill. Let's retrace that path in a bit more detail.

The Men Who Tamed Motion

When mathematicians in the sixteenth and early seventeenth century tried to write down formulas that describe the motions of the planets, they faced a fundamental problem. The tools of mathematics are essentially static. Numbers, points, lines, and so forth are fine for counting and measuring, but they do not, on their own, allow you to describe motion. In order to study continuously moving objects, they had to find a way to apply those static tools to study patterns of change. The key breakthrough was made independently by two mathematicians in the middle of the seventeenth century: Isaac Newton in England and Gottfried Leibniz in Germany. The method they developed is known nowadays as the differential calculus.

As I explained in Chapter 1, the method of calculus is similar to making a movie. If you take a sufficiently rapid sequence of still photographs of a moving scene and project them onto a screen at twenty-four frames a second or faster, the human brain will interpret the result as continuous motion. Newton and Leibniz's idea was likewise to regard continuous motion as made up of a sequence of still configurations. Each still configuration could be analyzed using existing mathematical techniques—principally geometry and algebra. The difficult part was putting all the still configurations together. Twenty-four frames a second will fool the human brain into thinking it is seeing continuous motion. To achieve continuous motion mathematically, Newton and Leibniz had to "project" the still configurations at infinite speed, and each configuration had to be infinitely short in duration. Differential calculus is the collection of techniques Newton and Leibniz (and later others) developed to perform this infinite sequencing. (See Figure 4.1.)

Isaac Newton was born on Christmas Day, 1642, in the Lincolnshire village of Woolsthorpe. In 1661, following a fairly normal grammar school education, he entered Trinity College, Cambridge, where, largely through self study, he acquired a mastery of astronomy and mathematics. In 1664 he was

Figure 4.1. In a movie, the continuous motion we see of a tennis player making a serve results from a rapid sequence of static images.

promoted to the rank of "scholar," a status that provided him with four years of financial support towards a master's degree.

In 1665, with London in the grip of the bubonic plague, the Cambridge authorities closed the university and sent the students and faculty home, as a precaution in case the deadly disease spread the fifty miles north from the capital. Newton returned to Woolsthorpe, whereupon he proceeded to change the course of civilization. At the age of just twenty-three, he embarked on two of the most productive years of original scientific thought the world has ever seen. The invention of the method of fluxions (his name for the differential calculus) was just one of several major accomplishments in mathematics and physics that he made during these years.

After returning to Cambridge in 1667, Newton completed his master's degree and a year later was elected a Fellow of Trinity College, a lifetime position. In 1669, when Isaac Barrow resigned the Lucasian Chair in order to become Chaplain to the King, Newton was appointed to the position.

An overwhelming fear of criticism kept Newton from publishing a great deal of his work, including the calculus, but in 1684 the astronomer Edmund Halley persuaded him to prepare for publication some of his work on the laws of motion and gravitation. The eventual appearance, in 1687, of *Philosophiae Naturalis Principia Mathematica* was to change physical science for all time, and established Newton's reputation as one of the most brilliant scientists the world has ever seen.

In 1696, Newton resigned his Cambridge chair to become Warden of the Royal Mint. It was while in charge of the British coinage that he published, in 1704, his book *Opticks*, a mammoth work outlining the optical theories he had been working on during his Cambridge days. In an appendix to this book he gave a brief account of the method of fluxions he had developed forty years previously. It was the first time he had published any of this work. A more thorough account, *De Analysi*, had been privately circulated among the British mathematical community from the early 1670s onward, but was not published until 1711.

A complete account of the calculus written by Newton did not appear until 1736, nine years after his death.

Just prior to the appearance of *Opticks*, Newton was elected President of the Royal Society, the ultimate scientific accolade in Great Britain. In 1705, Queen Anne bestowed on him a knighthood, the ultimate royal tribute. He died in 1727 at the age of 84 and was buried in Westminster Abbey. His epitaph in the Abbey reads, "Mortals, congratulate yourselves that so great a man has lived for the honor of the human race."

The other inventor of calculus, Gottfried Leibniz, was a philosopher's son, born in Leipzig in 1646. He was a child prodigy, who took full advantage of his father's sizable scholarly library. At fifteen years, the young Leibniz entered the University of Leipzig. Five years later he had completed his doctorate and was set to embark on an academic career. But then, suddenly, he decided to leave university life and enter government service.

In 1672, Leibniz became a high-level diplomat in Paris, from where he made a number of trips to Holland and Britain. These visits brought him into contact with many of the leading academics of the day, among them the Dutch scientist Christian Huygens, who inspired the young German diplomat to resume his studies in mathematics. It proved to be a fortuitous meeting. By 1676, Leibniz had progressed from being a virtual novice in mathematics to having discovered for himself the fundamental principles of the calculus.

Or had he? When Leibniz first published his findings in 1684, in a paper in the journal *Acta Eruditorum*, of which he was the editor, many British mathematicians cried foul, accusing Leibniz of taking his ideas from Newton. Certainly, on a visit to the Royal Society in London in 1673, Leibniz had seen some of Newton's unpublished work, and in 1676, in response to a request for further information about his discoveries, Newton had written two letters to his German counterpart, providing some of the details.

Though the two men themselves largely stayed out of the debate, the argument between the British and German mathematicians over who had invented the calculus grew heated. Certainly,

Newton's work had been carried out before Leibniz's, but the Englishman had not published any of it. In contrast, not only had Leibniz published his work promptly, but his more geometric approach led to a treatment that is in many ways more natural and that quickly caught on in Europe. Indeed, to this day, Leibniz's geometric approach to differentiation is the one generally adopted in calculus classes the world over, and Leibniz's notation for the derivative (dy/dx, as we shall see presently) is in widespread use, whereas Newton's approach in terms of physical motion and his notation are rarely used outside of physics. Today, the general opinion is that although Leibniz almost certainly obtained some of his ideas from Newton's work, the German's contribution was undoubtedly significant enough to give both men the credit for inventing the calculus.

Leibniz was no more content than Newton to spend his entire life working in mathematics. He studied philosophy; he developed a theory of formal logic, a forerunner of today's symbolic logic; and he became an expert in the Sanskrit language and the culture of China. In 1700, he was a major force in the creation of the Berlin Academy, which he served as president until his death in 1716. Unlike Newton, however, who was given a state funeral in Westminster Abbey, Germany's creator of the calculus was buried in quiet obscurity.

What Is Calculus?

So much the for the men who created calculus. What exactly is it?

The differential calculus provides a means to describe and analyze motion and change. Not just any motion or change. There has to be a pattern that describes its occurrence. In concrete terms, the differential calculus is a collection of techniques for the manipulation of patterns. (The word "calculus" is Latin for pebble. Early counting systems involved the physical manipulation of pebbles.)

The basic operation of the differential calculus is the process known as differentiation. The aim of differentiation is to obtain the rate of change of some changing quantity. In order to do this, the value or position or path of that quantity has to be given by an appropriate formula. Differentiation then acts upon that formula to produce another formula that gives the rate of change. Thus, differentiation is a process for turning formulas into other formulas.

For example, suppose a car travels along a road and that the distance traveled, say x, varies with the time, t, according to the formula

$$x = 5t^2 + 3t$$

Then, according to the differential calculus, the speed s (i.e., the rate of change of position) at any time t is given by the formula

$$s = 10t + 3$$

The formula $10t + 3$ is the result of differentiating the formula $5t^2 + 3t$. (You will see shortly just how differentiation works in this case.)

Notice that the speed of the car is not a constant in this example; it varies with time, just as does the distance. The process of differentiation may be applied a second time to obtain the acceleration (the rate of change of the speed). Differentiating the formula $10t + 3$ produces the acceleration

$$a = 10$$

which in this case is a constant.

The fundamental mathematical objects to which the process of differentiation applies are called functions. Without the notion of a function, there can be no calculus. Just as arithmetical addition is an operation that is performed on numbers, differentiation is an operation that is performed on functions.

But what exactly is a function? The simplest answer is that in mathematics, a function is a rule that given one number allows you to calculate another. (Strictly speaking, this is a special case, but it is adequate for understanding how the calculus works.)

For example, a polynomial formula such as

$$y = 5x^3 - 10x^2 + 6x + 1$$

determines a function. Given any value for x, the formula tells you how to compute a corresponding value for y. For instance, given the value $x = 2$, you may compute

$$y = (5 \times 2^3) - (10 \times 2^2) + (6 \times 2) + 1 = 40 - 40 + 12 + 1 = 13$$

Other examples are the trigonometric functions $y = \sin x$, $y = \cos x$, $y = \tan x$. For these functions there is no simple way to compute the value of y as we could in the case of a polynomial. Their familiar definitions are given in terms of ratios of the various sides of right-angled triangles, but those definitions apply only when the given x is an angle less than a right angle. The mathematician defines the tangent function in terms of the sine and cosine functions as

$$\tan x = \frac{\sin x}{\cos x}$$

and defines the sine and cosine functions by means of infinite sums:

$$\sin x = x - \frac{x^3}{3!} + \frac{x^5}{5!} - \frac{x^7}{7!} + \cdots$$

$$\cos x = 1 - \frac{x^2}{2!} + \frac{x^4}{4!} - \frac{x^6}{6!} + \cdots$$

To understand these formulas, you need to know that, as we saw in Chapter 3, for any natural number n, $n!$ (read "n-factorial") is equal to the product of all numbers from 1 to n inclusive. For example, $3! = 1 \times 2 \times 3 = 6$. You also need to understand that those three dots mean that the series continues in the same pattern to infinity. The infinite sums for $\sin x$ and $\cos x$ always give a finite value, and may be manipulated more or less like finite polynomials.

Still another example of a function is the exponential function

$$e^x = 1 + \frac{x^1}{1!} + \frac{x^2}{2!} + \frac{x^3}{3!} + \frac{x^4}{4!} + \cdots$$

Again, this infinite sum always gives a finite value, and may be manipulated like a finite polynomial. Putting $x = 1$, you get

$$e = e^1 = 1 + \frac{1}{1!} + \frac{1}{2!} + \frac{1}{3!} + \frac{1}{4!} + \cdots$$

The mathematical constant e that is the value of this infinite series is an irrational number. Its decimal expansion begins 2.71828.

The exponential function e^x has an important inverse function, that is to say, a function that exactly reverses the effect of e^x. It is the natural logarithm, $\ln x$, which we met in Chapter 2. If you start with a number a and apply the function e^x to get the number $b = e^a$, then, when you apply the function $\ln x$ to b you get a again: $a = \ln b$.

How to Compute Slopes: The Derivative

Algebraic formulas such as polynomials, or the infinite sums for the trigonometric or exponential functions, are a very precise way to describe a certain kind of abstract pattern. The pattern in these cases is a pattern of association between pairs of numbers: the independent variable or argument x that you start with and the dependent variable or value y that results. In many cases this pattern can be illustrated by means of a graph. The graph of a function shows at a glance how the variable y is related to the variable x.

For example, in the case of the sine function shown in Figure 4.2, as x increases from 0, y also increases, until somewhere near $x = 1.5$ (the exact point is $x = \pi/2$) y starts to decrease; y

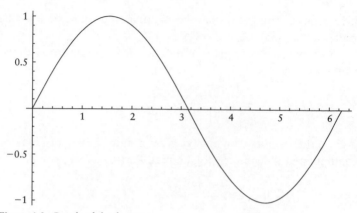

Figure 4.2. Graph of the function $y = \sin x$.

becomes negative around $x = 3.1$ (precisely, when $x = \pi$), continues to decrease until around $x = 4.7$ (precisely, $x = 3\pi/2$),
then starts to increase again.

The task facing Newton and Leibniz was this: How do you
find the rate of change of a function such as $\sin x$? That is, how
do you find the rate of change of y with respect to x? In terms
of the graph, this is the same as finding the slope of the curve—
how steep is it? The difficulty is that the slope is not constant;
at some points the curve is climbing fairly steeply (large positive slope), at other points the curve is almost horizontal (slope
close to zero), and still elsewhere the curve is falling fairly steeply
(large negative slope).

In summary, just as the value of y depends on the value of
x, so too the slope at any point depends on the value of x. In
other words, the slope of a function is itself a function, a second
function. The question now is, given a formula for a function—a
formula that describes the pattern relating x to y—can you find
a formula that describes the pattern relating x to the slope?

The method that both Newton and Leibniz came up with is,
in essence, as follows. (The approach I adopt is close to that of
Leibniz; Newton described things differently and used a somewhat different notation.) For simplicity, consider the function

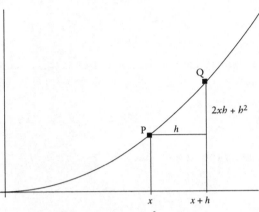

Figure 4.3. Derivative of the function $y = x^2$.

$y = x^2$, whose graph is shown in Figure 4.3. As x increases, not only does y increase, but the slope also increases. That is, as x increases, the curve not only climbs higher, it gets steeper. Given any value for x, the height of the curve for that value of x is given by computing x^2, but what do you do to x in order to compute the *slope* for that value of x?

The idea is this. Look at a second point a short distance h to the right of x. Referring to Figure 4.3, the height of the point P on the curve is x^2, and the height of Q is $(x + h)^2$. The curve bends upward as you go from P to Q, but if h is fairly small (as shown), the difference between the curve and the straight line joining P to Q is also small. So the slope of the curve at P will be close in value to the slope of this straight line.

The point of this move is that it is easy to compute the slope of a straight line: you just divide the increase in the height by the increase in the horizontal direction. In this case, the increase in the height is

$$(x + h)^2 - x^2$$

and the increase in the horizontal direction is h, so the slope of the straight line from P to Q is

$$\frac{(x+h)^2 - x^2}{h}$$

Using elementary algebra, the numerator in this fraction simplifies as follows:

$$(x+h)^2 - x^2 = x^2 + 2xh + h^2 - x^2 = 2xh + h^2$$

So the slope of the straight line PQ is

$$\frac{2xh + h^2}{h}$$

Canceling the h from this fraction leaves

$$2x + h$$

This is a formula for the slope of the straight line from P to Q. But what about the slope of the curve $y = x^2$ at the point P, which is what we started out trying to compute? This is where both Newton and Leibniz made their brilliant and decisive move. They argued as follows. Replace the static situation with a dynamic one, and think about what happens when the distance h that separates the two points P and Q along the x-direction is made smaller and smaller.

As h gets smaller, Q moves closer and closer to P, and for each value of h, the formula $2x+h$ gives the corresponding value for the slope of the straight line PQ. For instance, if you take $x = 5$ and let h successively assume each of the values 0.1, 0.01, 0.001, 0.0001, and so on, then the corresponding PQ slopes are 10.1, 10.01, 10.001, 10.0001, and so on. At once we see an obvious numerical pattern: The PQ slopes appear to be approaching the value 10.0. (In mathematician's jargon, 10.0 looks like it is the "limiting value" of the sequence of PQ slopes for smaller and smaller values of h.)

But by looking at the diagram and picturing this process geometrically, we can also see another pattern, a geometric one:

As h gets smaller and Q approaches P, the difference between the slope of the straight line PQ and the slope of the curve at P also gets smaller. The limiting value of the slope of PQ will be precisely the slope of the curve at P.

For instance, for the point $x = 5$, the slope of the curve at P will be 10.0. More generally, the slope of the curve at P for an arbitrary point x will be $2x$. That is to say, the slope of the curve at x is given by the formula $2x$ (which is the limiting value of the expression $2x + h$ as h approaches 0).

In his description of the method, Leibniz used the notation dx in place of our h, and dy to denote the the difference in height between P and Q. He denoted the slope function by $\dfrac{dy}{dx}$, a notation obviously suggestive of a ratio of two small increments. (The notation dx is normally read "dee-ex," dy is read "dee-wye," and $\dfrac{dy}{dx}$ is read "dee-wye by dee-ex," or simply "dee-wye dee-ex.")

Regardless of the notation, the important starting point is to have a functional relation linking the two quantities:

$$y = \text{some formula involving } x$$

In modern terminology, we say that y is *a function of* x, and use notation such as $y = f(x)$ or $y = g(x)$.

Notation aside, the crucial step made by both Newton and Leibniz was to shift attention from the essentially static situation concerning the slope at a particular point P to the dynamic process of successive approximation of the slope by slopes of straight lines starting from P. It was by observing numerical and geometric patterns in this process of approximation that Newton and Leibniz were able to arrive at the right answer.

Moreover, their approach works for a great many functions, not just for the simple example considered above. For example, if you start with the function x^3, you get the slope function $3x^2$. More generally, if you start with the function x^n, where n is any natural number, the slope function works out to be nx^{n-1}. With this, you have another easily recognizable, if somewhat unfamiliar, pattern, the pattern that takes x^n to nx^{n-1} for any value of n. This is a pattern of differentiation.

It should be stressed that what Newton and Leibniz were doing was not at all the same as setting the value of h equal to 0. True, in the case of the very simple example above, where the function is x^2, if you simply set $h = 0$ in the slope formula $2x + h$, then you obtain $2x$, which is the right answer. But this is because the example is a simple one. In the geometric picture, if $h = 0$, then the points Q and P are one and the same, so there is no straight line PQ. (Remember that, although a factor of h was canceled to obtain a simplified expression for the slope of PQ, this slope is the ratio of the two quantities $2xh + h^2$ and h, and if you put $h = 0$, then this ratio reduces to the division of 0 by 0, which is meaningless.)

In fact, although Newton and Leibniz got the right answer, it took two hundred years for mathematicians to fully explain why the method worked. To do this, they had to work out a rigorous mathematical theory of approximation processes—something neither Newton nor Leibniz was able to do. It was not until 1821 that Augustin-Louis Cauchy developed the key idea of a "limit" (of a varying quantity), and a few years later still that Karl Weierstrass provided a formal definition of the notion of a limit. Only then was the calculus placed on a sound footing.

The process of going from the formula for a curve to a formula for the slope of that curve is known as *differentiation*. (The name reflects the idea of taking small differences in the x and y directions and computing the slopes of the resulting straight lines.) The slope function is called the *derivative* of the initial function (from which it is "derived").

For the example we have been looking at, the function $2x$ is the derivative of the function x^2. Similarly, the derivative of the function x^3 is $3x^2$, and in general, the derivative of the function x^n for any natural number n is nx^{n-1}. Exhibiting the kind of symmetry that mathematics often yields, the derivative of $\sin x$ turns out to be $\cos x$, so $\cos x$ gives the slope of $\sin x$ at any point; almost as pleasing, the derivative of $\cos x$ is $-\sin x$, with only that minus sign breaking perfect symmetry. Even nicer, the derivative of e^x is itself e^x, which means that e^x gives its own slope at any point. It is the only function with that property. The derivative of $\ln x$ is $1/x$.

The power of Newton's and Leibniz's invention was that the number of functions that could be differentiated was greatly enlarged by the development of a *calculus*, a series of rules for differentiating complicated functions. The development of this calculus also accounts for the method's immediate, enormous success in different applications, despite its dependence on methods of reasoning that were not fully understood at the time. People knew what to do, even if they did not know why it worked. (Many students in today's calculus classes have a similar experience.)

Taking Calculus into a Higher Dimension

The calculus as I have described it so far applies to functions of one variable, i.e., functions that take a single variable x and produce a value y. Such functions can be represented geometrically by means of a two-dimensional graph, generally a curved line. But the same ideas can be made to work in more general situations. In particular, calculus can be developed for functions of two variables, $z = f(x, y)$, where the geometric interpretation is a surface (with z being the height of the surface above the xy-plane at the point (x, y)) (see Figure 1.3 on page 34) or for functions of three or more variables.

For functions of three variables, say $v = f(x, y, z)$, there is no simple geometric picture, but many familiar physical phenomena take this form. For example, (x, y, z) could be the coordinates of a helicopter in flight and $v = g(x, y, z)$ could be the helicopter's speed when it has latitude x, longitude y, and altitude z. Or $w = T(x, y, z)$ could be the temperature of the earth's atmosphere at latitude x, longitude y, and altitude z.

In the case of a function of two variables, there is no such thing as a single rate of change at a point. It depends on which direction you are going. For example, suppose you are halfway up a mountain. If you continue to head directly for the summit, the slope of your path will be a positive number, possibly a large one. But if you decide to stop climbing and simply circle the hill at your current altitude, the slope of your path is zero. In other words, the slope at your location depends on which way you wish to travel from that point. In the same way, for functions of three or more variables, the slope at any point depends on the direction of travel. For example, a helicopter might have zero vertical speed, high forward speed, and a small sideways speed to the right.

It turns out that to analyze motion in two or more dimensions, you don't have to compute rates of change (i.e., the slopes) in all possible directions. It's enough to know the rates of change in the direction of the axes—that is, the rates of change in the x- and the y-directions in the case of a function $z = f(x, y)$, and the rates of change in the x-, y-, and z-directions for a function $v = g(x, y, z)$.

For instance—and we'll soon make direct use of this example—suppose you wanted to describe the motion of a speck of dust carried along in a flowing fluid. The motion of the speck could be quite complicated, as it is thrust first in one direction, then another, sometimes moving in a fairly straight line, other times spiraling around as the current takes it along. At each moment in time, t, we can specify the exact position of the speck by giving its three coordinates x, y, z (for some fixed coordinate system). We can likewise specify its motion at time t by describing how it moves in each of the three coordinate directions, for example, calculating the rate of change of position in the x-direction, the rate of change of position in the y-direction, and the rate of change of position in the z-direction. (When we do this using calculus, these are called the "directional derivatives" in the x-, y-, and z-directions.)

This idea of describing motion in three-dimensional space by describing the motion in each of the three coordinate directions forms the basis of Euler's applications of calculus to the study of fluid motion, which Navier and Stokes later elaborated. Here is the general idea.

For simplicity, let's start with the case of a function $z = f(x, y)$ of two variables. The directional rates of change in the axial directions are calculated by taking the appropriate cross section of the geometric figure concerned (the surface in the case of a function $z = f(x, y)$) and applying the standard methods of single-variable calculus to that cross section.

To indicate that differentiation is carried out in a particular direction, mathematicians use a slightly modified notation. For a function $y = f(x)$, where there is just one independent variable, x, the rate of change of y at the point x is denoted by $\dfrac{dy}{dx}$. For a function $z = f(x, y)$, where there are two independent variables x, y, the rate of change of z in the x-direction is denoted by $\dfrac{\partial z}{\partial x}$, and the rate of change in the y-direction is denoted by $\dfrac{\partial z}{\partial y}$. The quantity (or formula) $\dfrac{\partial z}{\partial x}$ is called the *partial derivative* of z with respect to x, and $\dfrac{\partial z}{\partial y}$ is called the *partial derivative* of z with respect to y. Thus $\dfrac{\partial z}{\partial x}$ is the slope you encounter when you move along the surface parallel to the x-axis; $\dfrac{\partial z}{\partial y}$ is the slope you encounter when you move parallel to the y-axis.

For example, for the function

$$z = x^2 - 3xy - y^5$$

we get

$$\frac{\partial z}{\partial x} = 2x - 3y \quad \text{and} \quad \frac{\partial z}{\partial y} = 3x - 5y^4$$

Similarly, for a function $v = f(x, y, z)$, the motion or change in v at any point (x, y, z) can be determined by the three partial derivatives $\dfrac{\partial v}{\partial x}, \dfrac{\partial v}{\partial y}, \dfrac{\partial v}{\partial z}$, which give the rates of change of (the value of) the function in each of the three axial directions. In this case, there is no comforting geometric visualization of the function, of course.

All of this was worked out, by many mathematicians, in the years following Newton and Leibniz's invention of calculus. At which point, enter Daniel Bernoulli.

From Balls and Planets to the Flow of Liquids

Daniel Bernoulli came from a large family of very talented Swiss mathematicians who lived in the eighteenth century. His father, Johann, was a professor of mathematics at the University of Basel. Both father and son were highly influenced by the method of infinitesimal calculus, and both helped to develop the new technique. At the time, calculus was used to study the continuous motion of solid objects such as planets (as Newton conceived it) or the continuously changing slopes of continuous geometric graphs (as in Leibniz's framework). Daniel Bernoulli sought to apply the method to the continuous motion of fluids, which to a scientist means liquids or gases. On the face of it, this was a very different problem.

For Newton and Leibniz, the continuous motion analyzed was that of a single, discrete object (a planet or a particle for Newton, the point that traces out a graph or a surface for Leibniz). In the case of a flowing fluid, however, not only the motion but the material itself is continuous. The idea for how to approach this problem was obvious enough. Just as regular calculus regarded continuous motion as made up of infinitesimally small discrete jumps, infinitesimally close in time, so too Bernoulli regarded a continuous fluid as made up of infinitesimally small discrete regions (or "blobs"), infinitesimally close together, each of which could (in principle) be handled using the methods of Newton and Leibniz.

Another way to think of this is that the aim is to write down equations that describe the path of any speck of dust—an "infinitesimal point"—located at any particular point in the fluid. This requires handling two sets of infinitesimals.

First, the motion of each infinitesimal speck is regarded as a sequence of "still frames." This is standard calculus for the continuous motion of an object, where the motion is regarded as a temporal sequence of static situations.

Then there is the infinitesimal geometric variation between the path taken by one speck and the path followed by another speck infinitesimally close by.

The challenge was to handle both kinds of infinitesimals— the temporal ones for the motion of each speck and the geometric ones for the fluid—at the same time. It took Bernoulli most of his adult life, but he did it. In 1738 he published his results in his book *Hydrodynamics*. The key idea was to take the solution to be what is called a *vector field*. I won't go into the gory details of how mathematicians formally define a vector field. Intuitively, it is a function of three variables x, y, z that tells you the speed and direction of the fluid flow at any point (x, y, z) in the fluid.

Among the results Bernoulli established in his book is an equation showing that when a fluid flows over a surface, the pressure the fluid exerts on the surface decreases as the speed of flow increases. Why is this worth mentioning? Because Bernoulli's equation (as the result is known today) forms the basis of modern aircraft flight. Put simply, Bernoulli's equation is what keeps an airplane in the air.[1]

Euler built on Bernoulli's work to formulate equations that describe the motion of a frictionless fluid subject to given forces, but he was unable to solve those equations. Navier and Stokes then modified Euler's equations to allow for viscosity (i.e., fluid

---

1. Many books and articles describe this incorrectly in terms of the shape of the wing, claiming that this forces the air to travel farther, and hence faster, over the upper surface, and thereby create lift. Since airplanes can fly upside down, this theory is clearly false. In fact, the lift is created by the air moving over the entire surface, wing and fuselage, and it is not (primarily) the wing's shape that gives the lift, but the angle to the horizontal at which the fuselage and wing cut through the air, called the attack angle.

friction). The resulting equations are known as the Navier–Stokes equations.

Though these equations can be solved in the hypothetical two-dimensional case of an infinitely thin planar film of fluid, it is not known whether there is a solution in the (more realistic) three-dimensional case. Notice that the issue is not, Do we know what the solution is? It's more basic than that. We don't even know whether there is a solution!

Let's begin with Euler's equations for fluid motion, the equations that govern flow in a (hypothetical) frictionless fluid that extends to infinity in all directions.

We assume that each point $P = (x, y, z)$ in the fluid is subject to a force that varies with time. We can specify the force at P at time t by giving its values in each of the three axial directions: $f_x(x, y, z, t)$, $f_y(x, y, z, t)$, $f_z(x, y, z, t)$. (To win the Clay Prize, it is enough to solve the problem in the case where there is no externally applied force, that is, when each of f_x, f_y, f_z is zero at all locations and all times. But historically, the problem was formulated in the way I am presenting it.)

Let $p(x, y, z, t)$ be the pressure in the fluid at the point P at time t.

The motion of the fluid at point P at time t can be specified by giving its velocity in the three axial directions. Let $u_x(x, y, z, t)$ be the velocity of the fluid at P in the x-direction, $u_y(x, y, z, t)$ the velocity of the fluid at P in the y-direction, and $u_z(x, y, z, t)$ its velocity in the z-direction.

We assume that the fluid is incompressible. That is, when a force is applied to it, it may flow in some direction but it cannot be compressed; nor can it expand. This is expressed by the following equation:

$$\frac{\partial u_x}{\partial x} + \frac{\partial u_y}{\partial y} + \frac{\partial u_z}{\partial z} = 0 \tag{1}$$

The problem assumes that we know how the fluid is moving at the start, i.e., when $t = 0$. That is, we know $u_x(x, y, z, 0)$, $u_y(x, y, z, 0)$, and $u_z(x, y, z, 0)$ (as functions of x, y, and z).

Moreover, these initial functions are assumed to be well-behaved ones. (Exactly what this means is technical, but we don't require a definition in order to obtain an overall understanding of the problem. The precise formulation of the restriction is, however, relevant to the Millennium Prize statement of the Navier–Stokes Problem, so would-be problem solvers would need to know the exact statement.)

By applying Newton's law

$$Force = Mass \times Acceleration$$

to each point P in the fluid, Euler produced the following equations, which when combined with the incompressibility equation (1) above describe the motion of the fluid:

$$\frac{\partial u_x}{\partial t} + u_x \frac{\partial u_x}{\partial x} + u_y \frac{\partial u_x}{\partial y} + u_z \frac{\partial u_x}{\partial z} = f_x(x, y, z, t) - \frac{\partial p}{\partial x} \quad (2)$$

$$\frac{\partial u_y}{\partial t} + u_x \frac{\partial u_y}{\partial x} + u_y \frac{\partial u_y}{\partial y} + u_z \frac{\partial u_y}{\partial z} = f_y(x, y, z, t) - \frac{\partial p}{\partial y} \quad (3)$$

$$\frac{\partial u_z}{\partial t} + u_x \frac{\partial u_z}{\partial x} + u_y \frac{\partial u_z}{\partial y} + u_z \frac{\partial u_z}{\partial z} = f_z(x, y, z, t) - \frac{\partial p}{\partial z} \quad (4)$$

Equations (1) through (4) are Euler's equations for fluid motion. To allow for viscosity, Navier and Stokes introduced a positive constant v, the *viscosity*, which measures the frictional forces within the fluid, and added an additional force—the viscous force—to the right-hand side of equations (2), (3), and (4). The term to be added to the right-hand side of equation (2) is

$$v \left[\frac{\partial^2 u_x}{\partial x^2} + \frac{\partial^2 u_x}{\partial y^2} + \frac{\partial^2 u_x}{\partial z^2} \right]$$

with entirely similar terms (with u_x replaced by u_y and u_z, respectively) added to equations (3) and (4).

Here, the notation $\frac{\partial^2 u_x}{\partial x^2}$ denotes the *second partial derivative*, obtained by first differentiating u_x with respect to x and then differentiating the result again with respect to x, i.e.,

$$\frac{\partial^2 u_x}{\partial x^2} = \frac{\partial}{\partial x}\left(\frac{\partial u_x}{\partial x}\right)$$

with analogous definitions in the y and z cases.

Unless you are a calculus whiz, chances are that the above formulas look pretty daunting. To be honest, mathematicians find them a bit overwhelming as well. The problem is that when we try to capture the motion of the fluid at any point in terms of its motion in each of the x-, y-, and z-directions, we make life unnecessarily complicated for ourselves. As you can see, there is relatively little difference among equations (2), (3), and (4), and the three additional viscosity terms we add are all variations on a single theme, one for each axial direction.

During the nineteenth century, mathematicians developed a notation and a method to handle directional motion in a simpler fashion. The idea was to introduce a new kind of quantity called a *vector*. Whereas a number simply has quantity, a vector has both quantity and direction. "Vector calculus" is the method you get when you develop calculus for vectors and vector functions instead of number-variables and number-variable functions. Using vectors, mathematicians can write the Navier–Stokes equations more compactly:

$$\frac{\partial \mathbf{u}}{\partial t} + (\mathbf{u} \cdot \nabla)\mathbf{u} = \nu \Delta \mathbf{u} - \text{grad } p, \qquad \text{div } \mathbf{u} = 0$$

Here, the quantity $\mathbf{u}$ is a vector function, and the symbols/terms ∇, Δ, grad, and div denote operations of vector calculus. (If you want to know more about this, consult one of the references given at the end of the chapter.)

So little progress has been made toward solving the Navier–Stokes equations that the Clay Institute will award the \$1 million prize for the solution to any one of several variations of the problem. The simplest version to state, though not necessarily the easiest to solve, assumes that you make the force functions f_x, f_y, and f_z all zero. Can you then find functions $p(x, y, z, t)$, $u_x(x, y, z, t)$, $u_y(x, y, z, t)$, and $u_z(x, y, z, t)$ that satisfy the modified versions of equations (1) through (4) (i.e., the versions that include the viscosity terms, with $v > 0$) and are sufficiently "well behaved" that they could plausibly correspond to physical reality?

Let me mention that the analogous problem where the viscocity is 0 (i.e., for the Euler equations) has also not been solved, but that version is not a Millennium Problem.

If the Navier–Stokes problem is reduced to two dimensions (by setting all z terms equal to 0), it can be solved. This is an old result, but it provides no clue to what happens in three dimensions.

The full three-dimensional problem can also be solved in a highly restricted way. Given the various initial conditions, it is always possible to find a positive number T such that the equations can be solved for all times $0 \leq t \leq T$. In general, the number T is fairly small, so this answer is not particularly useful in real life. The number T is called the "blowup" time for the particular system.

Will the Navier–Stokes Problem Be Solved?

Will anyone win the $1 million prize for solving the Navier–Stokes equations? By choosing this challenge as a Millennium Problem, the Clay Institute has shined the mathematical spotlight onto a part of mathematics that goes back over two hundred years: the calculus of fluid flow. Given the length of time that mathematicians have been trying to solve these equations, it is hard to deny the thought that they may simply be unsolvable. At the very least, it seems likely that a solution will require some genuine new techniques. The equations many look like a problem in a typical student textbook. But they are very definitely far more difficult than that.

FIVE

The Mathematics of Smooth Behavior

The Poincaré Conjecture

Henri Poincaré, who formulated what is now the sixth Millennium Problem, was born in Nancy, France, in 1854. The Poincarés were, to say the least, a family of high achievers. Henri's father, Léon, was a professor of medicine, and one of his cousins, Raymond, was prime minister of France several times and president of the French Republic during World War I. For his part, Henri became one of the greatest and most innovative mathematicians and physicists the world has ever seen. He came very close to beating Einstein to the discovery of special relativity, although not quite close enough to get the credit, but still achieved a place in the history books for (among many other things) almost single-handedly creating the enormously important branch of modern mathematics called algebraic topology—of which more presently. Because of the enormous span of his knowledge and achievements, which ranged over several branches of mathematics, over celestial mechanics, modern physics, and even psychology, Poincaré has been called the last great universalist of science.

Like Riemann, whose concept-based approach to mathematics he adopted, Poincaré suffered from poor health as a child. He was nearsighted, had poor muscular coordination, and at one time was seriously ill with diphtheria. Unlike Riemann, however, who was something of a problem pupil and never achieved fluent use of his native language, Poincaré excelled in all subjects apart from art and physical exercise, and even in elementary school he demonstrated a mastery of written composition that in adulthood would make him a world renowned expositor of science, with such popular science books as *Science and Hypothesis* (1901), *The Value of Science* (1905), and *Science and Method* (1908).

From 1862 to 1873, Poincaré attended the lycée (secondary school) in Nancy, now renamed the Lycée Henri Poincaré in his honor, where he won several national scholastic prizes. From high school he went to the prestigious Ecole Polytechnique in Paris, where his professors reported that he exhibited an awesome memory, not merely for rote learning but for understanding what he learned in a deep way. He excelled at linking together new ideas, often in a visual fashion.

Poincaré's preference for visual thinking was characteristic of much of the mathematics he produced throughout his career. (Some historians have speculated that because of his poor eyesight, he was often unable to see what his professors wrote on the board, and as a result he had to create his own pictures in his head, thereby strengthening his visualization skills.) Throughout his professional life, Poincaré, much like Riemann, preferred to work things out for himself from first principles rather than build on others' results or even on his own previous work.

After graduating from the Ecole Polytechnique in 1875, Poincaré went on to the Ecole des Mines, and then to a job as a mining engineer at Vesoul. Despite an interest in all aspects of mining that would continue throughout his life, by then he knew that his main passion was mathematics. While working in Vesoul, he wrote a doctoral dissertation on differential equations under Charles Hermite at the University of Paris. After his doctorate was awarded, in 1879, he took a teaching position at the University of Caen, but just two years later he was appointed to a chair in Paris—his awesome talent as a mathematician more than compensating for what by all accounts was an uninspiring and disorganized lecturing style. In 1886, he was appointed to the chair of mathematical physics and probability theory at the Sorbonne, which he held along with a chair at the Ecole Polytechnique until his death in 1912 at the early age of 58.

In addition to his powers as a mathematician—most mathematicians today rank him as one of the greatest geniuses of all time—Poincaré was a superb writer of what would today be called "popular science." He also had a deep interest in the nature of mathematical thought. In addition to giving a celebrated lecture on mathematical creativity at the Institut Général Psychologique in Paris in 1908, titled *Mathematical Invention* and based on introspection into his own thought processes, he also collaborated with Edouard Toulouse, the director of the Psychology Laboratory of the Ecole des Hautes Etudes in Paris, as part of the latter's studies of the working habits of very accomplished people. Toulouse published his results on the mathematician in 1910, in a book titled simply *Henri Poincaré*.

According to Toulouse, Poincaré followed a strict daily schedule. He did mathematical research between 10 and noon each morning and then again from 5 to 7 in the afternoon. Later in the evening he might read a journal article he wanted to know about, but apart from that, he avoided all serious work in the evenings. He believed that a mathematically trained brain works on mathematical problems subconsciously during sleep, so he did everything he could to ensure an untroubled night's rest.

Toulouse also tells us that when Poincaré was in the middle of a problem, it was virtually impossible to distract him, but if he reached a point where he did not know how to proceed, he would break off and do something else—convinced that his subconscious mind would continue to mull the problem over.

Poincaré himself wrote, "It is by logic we prove, it is by intuition that we invent."[1] He was particularly dismissive of

---

1. *Mathematical definitions in education*, 1904.

Hilbert's view that mathematical deduction could be axiomatized and (in principle) "mechanized," a program that Poincaré thought could not possibly succeed. (As we saw in Chapter 3, Gödel was subsequently to prove him right.)

The Last Universalist

Poincaré's research interests spanned many areas of mathematics, physics, and the philosophy of science, and he is the only person ever elected to all five sections of the French Académie des Sciences: geometry, mechanics, physics, geography, and navigation. He was also president of the Académie, in 1906. The wide range of his knowledge and his ability to see connections between seemingly very distinct areas allowed him to attack problems from many different and often novel angles. His work in physics included significant contributions to optics, electricity, telegraphy, elasticity, thermodynamics, potential theory, celestial mechanics, cosmology, fluid mechanics, quantum theory, and the theory of special relativity.

One of his first major contributions to mathematics, made while he was still in his twenties, was his development of the concept and theory of what are today called automorphic functions, a special class of functions from complex numbers to complex numbers. (See Chapter 1 for a discussion of functions and complex numbers.) These functions arose from a particular set of problems Poincaré had struggled with in his youth. Later, in *Science and Method*, he described how, after struggling with the problem for some time without success, he was stepping onto a bus one day, not consciously thinking about mathematics, when suddenly the key idea came to him that enabled him to define the new class of functions.

Later in his career, Poincaré did further work on functions involving complex numbers, and he is generally credited as the originator of the hugely important theory of analytic functions of several complex variables. At various stages of his life he also applied his talents to the study of both number theory and geometry.

But it is Poincaré's work in the branch of mathematics called topology that is of interest to us here. It is in topology that the sixth Millennium Problem arises: the Poincaré Conjecture. Although topology has its origins with the work of Gauss and others in the mid-nineteenth century, it really began in earnest only in 1895, when Poincaré published his book *Analysis situs* (the analysis of position). In that single publication, Poincaré introduced practically all the concepts and key methods that would drive the subject for the next fifty years.

Topology is a kind of "ultra geometry," which grew out of ordinary geometry and calculus, in which the mathematician studies very general properties of surfaces and other mathematical objects. One of Poincaré's major contributions was to invent ways to apply algebraic techniques to facilitate the study. The Poincaré Conjecture arose by accident, as a result of a mistake Poincaré made (and quickly noticed) right at the start of his investigation of this new geometry. Much of the interest in topology focuses on mathematical objects of three or more dimensions, and Poincaré's mistake was to assume that a fairly obvious fact about two-dimensional objects was also true for analogous objects of three or more dimensions.

To understand this mistake, and to see what the Poincaré Conjecture says, it's probably best to start with a brief discussion of two-dimensional topology, and then see what is involved when you try to move up to higher dimensions.

Two-dimensional topology is sometimes called, rather suggestively, "rubber-sheet geometry."

Rubber-Sheet Geometry

Anyone who has visited London—and many who haven't—will recognize the map shown in Figure 5.1. It's the standard map of the London Underground, which you will find posted all over the London subway system as well as adorning souvenir T-shirts, coffee mugs, and breakfast trays. Designed in 1931 by Henry Beck, a twenty-nine-year-old temporary draughtsman working for the London Underground System, it is widely regarded as one

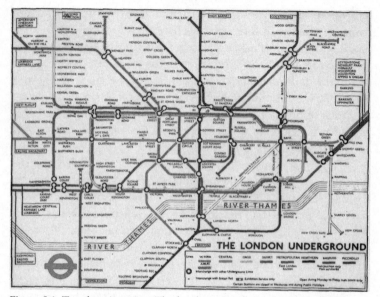

Figure 5.1. Topology in action: The familiar map of the London Underground.

of the best maps ever drawn, and several attempts to improve it have all met with failure. Somehow, the map manages to combine ease of use with an aesthetic overall appearance, making it an easily recognized icon of present-day London as well as a model for subway maps worldwide.

Yet how many who use the map realize that it illustrates the immense power of topology? It does so because in every respect save two, the map is completely inaccurate. It is not drawn to scale, and as a result the distances are all wrong. What's more, those neatly drawn straight lines representing the trains' paths bear little resemblance to the actual subway lines, in which the trains threaten to throw standing passengers to the floor as they twist and turn their way beneath the London streets. And just because a stretch of line is shown as running north–south, it doesn't mean the real line does the same—it could even run almost east–west. One of the two things that the map does get right is that if a station is shown to be north of the River Thames, then the real station is indeed north of the Thames, and likewise for south. The other accurate feature of the map is the way it depicts the network: the order in which the stations lie on each line and the places (stations) where any two lines intersect.

This property is really the only item of information an Underground traveler needs to know from the map—where to get on and off, and where to change lines. The Underground map works by being completely accurate in its depiction of the one thing travelers need to know to use the system, and sacrificing all other details in favor of a clear and attractive design.

Similarly, the wiring diagram for a supermarket refrigeration unit shown in Figure 5.2 doesn't tell you how long each length of wire should be or where it should be laid; it simply shows how the components should be connected together—the configuration of the network. Again, the diagram works by accurately

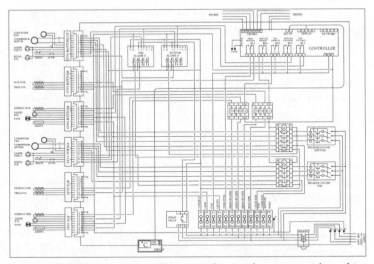

Figure 5.2. Topology in action. This wiring diagram for a supermarket refrigeration unit shows how the different components should be wired together, but does not specify the exact layout of the components or the lengths of the wires and the paths they take. (Drawing courtesy of R. T. Carey.)

portraying the one piece of information the engineer needs in order to construct the device, sacrificing all other details for a clear picture.

Both of these examples illustrate the essence of two-dimensional topology. If the Underground map were printed on a perfectly elastic sheet of rubber, it could be stretched and compressed so that every detail was correct, giving a standard, geographically accurate map, drawn to scale, with every stretch of line correctly oriented to the compass bearings. This stretching would not affect the way the lines connect the various stations. The reason, in mathematical terms, is that the configuration of a network (defined as a collection of points connected by various lines) is a *topological* property. Put simply, networks are topological objects. You can twist or stretch any of the connecting lines in a network without changing the overall configuration. To change the network, you must either break a connection or add a new one.

What holds for the London Underground map holds for any network. It holds for electrical circuit diagrams, the circuits themselves, computer chips, telephone networks, and the Internet. This is why "rubber-sheet geometry" is one of the most important branches of mathematics in the world today. In the case of the Underground map, as long as it is *topologically* accurate, the exact design does not matter. Similarly, when it comes to the design of an electrical circuit or a computer chip, what counts is the configuration of the network. Provided that it is *topologically* accurate, the exact layout of the wires or the conductive channels in the silicon can be altered at will to make the resulting device as fast, efficient, reliable, and (these days) as small as possible.

In general, two-dimensional topology (rubber-sheet geometry) studies the properties of figures that are preserved when drawn on a (hypothetical) perfectly elastic sheet of rubber that is then twisted and stretched. As we shall see, network configurations are just one of several properties that are not affected by such manipulations. In fact, although maps have always been important, and although the mathematics of networks is important in today's world, these were not the applications that motivated the original study of topology. Indeed, the development of topology was not driven by the needs of any area of applied mathematics. Rather, it came from within pure mathematics itself, from the struggle to understand why the differential calculus worked.

Understanding Mathematical Magic

Toward the end of the nineteenth century, mathematicians began to take a very close look at the assumptions—sometimes explicit but more often implicit—that underlay their subject. Much of the motivation for doing so came from attempts to understand how calculus worked. From the moment Newton and Leibniz introduced calculus in the middle of the seventeenth century, mathematicians used it both extensively and successfully. Put simply, calculus worked. And yet no one really understood why. It was a kind of magic.

The mathematical explanation of how calculus works was provided by the cumulative efforts of a great many mathematicians over a three-hundred-year period. To come up with that explanation, they had to carry out a detailed analysis of the nature of the real numbers, of infinite processes (such as the infinite sums and products we met in Chapter 1), and mathematical reasoning itself.

The drive for ever more detailed analysis was accompanied by—and to a large extent driven by—a dramatic increase in abstraction. As we saw in Chapter 1, in the middle of the nineteenth century, mathematics underwent something of a revolution. Since that time, the subject has become increasingly abstract.[2] For most of its history, mathematics dealt with objects and patterns that came from our everyday experience. Arithmetic deals with numbers, which, though technically abstract, are part of the fabric of our lives. Even real numbers, for all their

---

2. In my book *The Math Gene*, I argue that coping with abstraction is the single biggest obstacle to anyone being able to do well at mathematics.

difficulties, arise from the intuitively simple idea of a continuous straight line. Geometry deals with idealized versions of the shapes we see every day. Probability theory examines random events, familiar to anyone who has tossed a coin or played with cards or dice. And while the symbols and equations of algebra might seem abstract to the nonmathematician, before the late eighteenth century those algebraic symbols generally denoted numbers, so the appearance of abstraction was essentially a linguistic illusion.

The nineteenth century, on the other hand, saw the appearance of a host of new kinds of objects and patterns that were definitely not part of everyday experience—or, more accurately, were not recognized as such. (Since most of them came from a careful analysis of existing mathematics, arguably they were part of everyday life, but a hidden, skeletal part.) Among the new objects and patterns studied by mathematicians over the last hundred and fifty years are geometries in which parallel lines meet (called "non-Euclidean geometries"), geometries of four and more dimensions, geometries of infinitely many dimensions, algebra where the symbols stand for symmetries in figures (called "group theory"), algebra where the symbols stand for logical thoughts ("propositional logic"), and algebra where the symbols stand for motions in two- or three-dimensional space ("vector algebra").

The development of topology was part of this proliferation of new abstractions. The idea was to develop a "geometry" that studies properties of figures that are not destroyed by continuous deformation, and thus do not depend on notions such as straight lines, circles, cubes, and so on, or on measurement of lengths, areas, volumes, or angles. The objects studied in topology were called topological spaces (just as geometry can be said to study geometric spaces—such as the familiar two-dimensional Euclidean space of high school geometry).

The connection between topology and the effort to understand how calculus works is fairly subtle. In essence, both depend on being able to handle the infinitely small. We saw in Chapter 4 why this is so in the case of calculus. But what do topological transformations have to do with the infinitely small? The answer is, everything. Indeed, they provide the key to truly coming to grips with the infinitely small. The point is this: Intuitively, the essence of a topological transformation is that two points that are "infinitely close" before the transformation remain "infinitely close" afterwards. (In a moment I'll explain why those quotation marks appear in this last sentence.) In particular, neither stretching, compressing, nor twisting a sheet of rubber will destroy closeness. Two points that start out close will remain close after the manipulation is completed.

You have to be a bit careful here. The notion of closeness involved is that of closeness relative to all the other points in the topological space. We can stretch the sheet so that two points initially close together no longer appear to us to be close together. But in this case the change in "closeness" is a geometric one that we impose from outside. From the rubber sheet's perspective, the two points remain close together. The only way to destroy that closeness is to cut or tear the sheet—a forbidden operation in topology.

To get topology off the ground, mathematicians had to find a way to get at that key idea of relative closeness. They set out to do this by looking for a way to capture the hypothetical notion of two points being "infinitely close." Intuitively, a topological transformation has the property that if two points start out infinitely close together, then they remain so after the transformation is carried out. The problem with that approach was that the notion of "infinitely close" was not a well-defined notion.[3] However, by thinking of topological transformations in this way, they were able to find a way to give a precise definition of a topological transformation. (It would be too much of a digression to explain what that definition is.) At that point,

---

3. To be accurate, when Poincaré and others were carrying out this work at the end of the nineteenth century, no one knew how to make the idea of "infinitely close" mathematically precise. In the 1950s, the American mathematician Abraham Robinson found a way to do so. But that does not affect our story.

they were able to turn the original analysis on its head, as it were, and use the concept of topological transformations (now properly defined and well understood) to analyze, in a precise way, the intuitive notion of "infinitely close" they had started out with. In this way, they were able to develop calculus in a rigorous way, avoiding the problematic concept of "infinitely close," instead using topology as an underlying basis.

And that is the main reason why Poincaré and others developed topology.

An obvious question that strikes anyone who meets topology for the first time is whether anything interesting can be said about topological spaces. After all, once you start throwing away important features such as straightness of lines or flatness of planes, there's no guarantee that you'll end up with anything of any real value. Topological spaces not only have no straight lines, they have no notion of a fixed shape at all, nor is there any kind of distance. All you can say is when two points are close to each other. That may be fine for putting calculus on a firm footing, but how far can it get you within topology itself?

There's More There Than You'd Think

The example of the London Underground map shows that topology isn't an entirely vacuous subject. Surprisingly, however, it turned out to be one of the richest and most fascinating and important branches of contemporary mathematics, having many applications in mathematics, physics, and other walks of life. To mention just one important application, topology forms the mathematical foundation of superstring theory, physicists' most recent theory of the nature of the universe.

Let's take a look at some of the things topologists study. For simplicity, I'll stay with the two-dimensional case (rubber-sheet geometry), and ask what properties from ordinary high school geometry carry over to topology. Since stretching and twisting the rubber will change straight lines into curves and will alter distances and angles, none of these familiar geometric concepts has any significance in topology. So what is left?

We still have lines, provided we don't demand that they be straight or circular or have any particular shape. What else? How about closed loops—lines that close back on themselves? If you draw a loop on a perfectly elastic sheet of rubber, then no matter how much you stretch, compress, or twist the rubber, the loop will remain a loop. What else?

To answer that, I'll show you the very first mathematical result that can justifiably be called a topological theorem. It is due to the great Swiss mathematician Leonhard Euler, whom we met in Chapter 1 as the inventor of the zeta function. In 1735, he solved a long-standing puzzle called the Königsberg bridges problem.

The city of Königsberg, which was located on the River Pregel in East Prussia—it is nowadays called Kaliningrad and is in Russia—had two islands, joined together by a bridge. One island was connected to each bank by a single bridge, while the other island had two bridges to each bank. Figure 5.3 gives a sketch of the city with its islands and bridges.

Many of the citizens of Königsberg used to take a family walk each Sunday, and, naturally enough, their path would often take them over several of the bridges. A question that was often discussed was whether there was a route that traversed each bridge exactly once.

Euler solved the problem by realizing that the exact layout of the islands and bridges was irrelevant. What was important was the way the bridges connect, that is to say, the network formed by the bridges. In other words, the problem was one of topology, not geometry. The problem remains the same if the island–bridge network is drawn more simply, as in Figure 5.4. The simplified diagram shown there has just four points (usually called "nodes" or "vertices" of the network) connected by seven lines (usually called the "edges" of the network).

Euler now argued as follows. Take any network and suppose you have a tour that traverses each edge exactly once. Any node that is not a starting or a finishing point of the tour must have an even number of edges meeting there, since those edges can be paired off into path-in–path-out pairs. But in the bridges

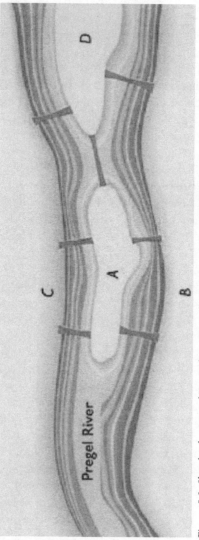

Figure 5.3. Sketch of a map of Königsberg, now the Russian city of Kaliningrad, showing the seven bridges that led to the problem eventually solved by Euler in 1735.

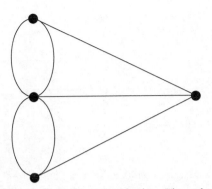

Figure 5.4. The network of the Königsberg bridges. The nodes of the network (the points) represent the land; the edges (lines) represent the bridges.

network, all four vertices have an odd number of vertices that meet there. Hence there can be no such path. In consequence, there can be no tour of the bridges of Königsberg that crosses each bridge exactly once.

The solution to the Königsberg bridges problem was Euler's first known theorem in topology—indeed, the world's first topological theorem—but it was not his last or his most significant. That accolade almost certainly goes to a remarkable topological result he discovered about networks. Euler showed that for any network drawn on a flat surface, if V is the total number of vertices (or nodes), if E is the total number of edges (or connecting lines), and if F is the total number of "faces" (i.e., regions enclosed by three or more edges), then the following simple formula is true:

$$V - E + F = 1$$

For example, Euler's own network for the Königsberg bridges has $V = 4$, $E = 7$, and $F = 4$, so

$$V - E + F = 4 - 7 + 4 = 1$$

For another example, look at the simple network in Figure 5.5. For this network, $V = 7$, $E = 10$, and $F = 4$, so

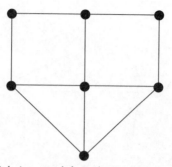

Figure 5.5. Proof of Euler's network formula. For this network, $V = 7$, $E = 10$, $F = 4$. Thus $V - E + F = 1$.

$$V - E + F = 7 - 10 + 4 = 1$$

Remarkably, the expression $V - E + F$ works out to be 1 for every network that has even been written down, and every one that ever could be written down. Euler gave a fairly simple proof of this fact. The general idea is to start with any network and, one by one, eliminate edges and end-nodes (nodes that have only one edge going to them). Removing an edge that does not go to an end-node decreases both E and F by 1 and leaves V unchanged, so the value of $V - E + F$ is unchanged. If you remove an end-node, you also remove the edge that goes into it. Removing an edge to an end-node does not change F, but it decreases both V and E by 1, so again $V - E + F$ is not changed. This process will end when all you have left is a single node. For this trivial network, $V = 1$, $E = 0$, and $F = 0$, so $V - E + F = 1$. But the value of $V - E + F$ remained the same throughout the entire reduction process. Hence the original value of $V - E + F$ for the network we started with must have been 1.

Although Euler solved one of the first topological puzzles and proved one of the first topological theorems, topology did not really get under way until the late nineteenth century, when Poincaré came onto the scene.

Getting Beneath the Surface

Here is the story so far. In topology we study properties of figures and objects that remain unchanged when the figure or object is subjected to a continuous deformation. By continuous, we mean that the deformation does not involve any cutting, tearing, or gluing. (Mathematicians often exclude, in addition, the creation or destruction of any sharp points or creases. I shall adopt this more stringent notion for my brief account of topology.)

For example, in topology, an (American) football is the same as a soccer ball, and both are the same as a tennis ball, since any one of these three kinds of ball can be continuously deformed into either one of the others. Another way of saying this is that in topology there is just one kind of "ball." The distinctions we normally make between different kinds of balls all have to do with size and shape, and these are not topological properties.

There is an old quip that a topologist is someone who cannot tell the difference between a coffee cup and a doughnut. (See Figure 5.6.) Imagine a doughnut made of soft modeling clay. You can manipulate the clay to turn the doughnut shape into a coffee cup (with a handle). The ring of the doughnut provides the handle, and you push most of the clay around the ring so you can fashion the body of the cup. This is all but impossible to do well with real clay, but with the perfectly elastic clay of the mathematician's imagination, it works just fine.

Incidentally, to avoid the temptation resulting from continuously talking of doughnuts, mathematicians—whose sedentary workstyle makes them prone to overweight and whose love of coffee prompted the late Hungarian mathematician Paul Erdős to quip, "A mathematician is a machine for turning coffee into theorems"—refer to the familiar ring shape of a doughnut as a torus.

As you might well expect, in a field where a coffee cup is the same as a doughnut, much of the early work in topology concerned a search for ways to tell when two shapes are topologically different. Poincaré himself was one of the leaders in this quest.

Figure 5.6. A doughnut and a coffee cup are topologically the same. One can be transformed continuously into the other. *Source:* Devlin: *The Language of Mathematics*, W. H. Freeman, Figure 6.13, page 239.

For instance, although any two balls are topologically the same, as are any two torus shapes (be they circular, oval-shaped, or whatever), any ball is topologically distinct from any torus. Intuitively, this seems clear. After all, there seems to be no way you could continuously transform a sphere to give a torus. The problem lies in that innocuous word "seems." How do you know for sure there is no way of doing it? Just because you have tried for an hour or so and not found the right sequence of moves doesn't mean there isn't one. For instance, can you find a way of continuously transforming figure (a) into figure (b) in the ring puzzle shown in Figure 5.7? The obvious way is to cut one of the interlocking rings, as shown in (c), separate the two rings, and then glue the cut ring back together again. But it can be done without cutting. The answer, perhaps surprisingly, is yes. Trying to see how to do it should convince you that the search for totally reliable ways of showing whether two objects are or are not topologically the same was an important task. (To avoid being deluged with letters asking me for the solution, I have given the correct sequence of moves at the end of the chapter.)

To repeat, simply not being able to find a continuous transformation that turns one object into another does not provide conclusive proof that the two objects are topologically different.

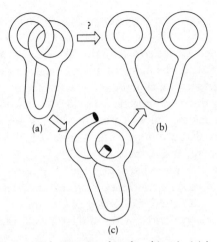

Figure 5.7. The ring puzzle. Imagine that the object in (a) is made of a perfectly elastic material. Can you transform it so that the two rings are no longer linked, as in (b)? The obvious way is to cut one of the rings, as in (c), separate the two rings, and then rejoin the cut ends. Provided the two cut ends are joined together exactly as before, this is an allowable topological operation from (a) into (b). However, it is possible to make the transformation without any cutting. Can you see how? The solution is given at the end of the chapter. *Source:* Devlin: *Mathematics: The New Golden Age*, Columbia University Press, Figure 48, page 226.

What is required is to find some topological property—that is, a property that is not changed by continuous deformation—that one of the two objects has and the other does not.

We have already met one such property. As we saw earlier, the value of the quantity $V - E + F$ for any network is a topological property. The quantity is the same for any network. Moreover, continuous deformation of a surface on which the network is drawn does not change the way the network is connected together, and hence does not alter any of the values V, E, and F. For the case we considered, namely, networks drawn on a plane, the value of $V - E + F$ works out to be 1. (Since this is the value on a plane, it follows that this is the value for a network drawn on any sheet, no matter how much it twists and turns.) If you consider networks drawn on the surface of a sphere (i.e., covering the entire sphere, not just one part of it), then $V - E + F = 2$. For networks drawn on a torus (again, drawn to cover the entire torus), then $V - E + F = 0$. Thus, we can conclude with absolute confidence that a two-dimensional plane, the surface of a sphere, and the surface of a torus are all topologically different. For networks drawn on a double-ringed torus (which has the shape of a figure eight), $V - E + F = -2$, so we also know that a double torus is topologically distinct from a plane, a sphere, and a torus.

Of course, for these particular four surfaces, it really is obvious that none can be continuously transformed into any of the others. But as our interlocking ring puzzle shows, once you get even a little way beyond spheres and toruses, things are far less obvious.

The value of the expression $V - E + F$ for any network drawn on a particular surface is an example of what mathematicians call a topological invariant of the surface. What this means is that this value remains invariant if we subject the surface to a topological transformation (i.e., to a continuous deformation). In honor of the man who first showed that this quantity is the same for any network drawn on a flat plane, the value of $V - E + F$ is called the Euler characteristic of the surface. It is one of a number of topological invariants that topologists have found that can be used to determine whether or not two particular surfaces are topologically equivalent.

Another topological invariant is the chromatic number of a surface. This has its origins in a classic problem about the coloring of maps. In 1852, a young English mathematician called Francis Guthrie asked the following seemingly innocuous question: What is the minimum number of colors you need in order to be able to color in the regions of any map? The one stipulation is that any two regions that share a stretch of common border have to be colored differently. (If two regions touch at a single point, that is not considered a common border.) It is very easy to draw maps that require four different colors, but are there maps that need five colors? The answer is no, but it took over a hundred years to prove this, and when a proof was finally produced in 1976, it involved not just sophisticated mathematical reasoning but also significant use of a computer. In fact, the Four Color Theorem, as it became known, was the first major mathematical result in which the use of a computer was considered unavoidable.

The Four Color Theorem is clearly a topological result, since continuous deformation of the sheet on which a map is drawn will not alter the pattern of shared borders. Two regions that share a stretch of common border before the deformation will do so afterward, and vice versa. Thus a proper coloring of the map before the deformation will still be a proper coloring afterwards.

The Four Color Theorem, like the original question it answered, is about maps drawn on a plane. But you can ask the same question of maps drawn on any surface. The chromatic number of a surface is the least number of colors you need to be able to color any map drawn on that surface. By the Four Color Theorem, the chromatic number of a plane is 4. So too is the chromatic number of (the surface of) a sphere. (The proof of the Four Color Theorem works for maps drawn on a plane or on a sphere.) The chromatic number of a torus is 7.

Taking Sides

Another topological invariant has its origins in the notion of "sidedness"—whether a surface has one or two sides. At first, this sounds silly. Surely, any surface has two sides, doesn't it? The answer is no. It is easy to construct a surface that has only one side. Take a long, thin strip of paper, say 1 inch wide and a foot long, give it a half twist, and then paste together the two short ends to form a twisted band, as shown in Figure 5.8. The twisted band is a surface that has only one side. You can check this by taking a pencil and drawing a line around the loop down the middle of the band. You will discover that the line goes around the loop twice and then comes back to the starting point. Since a line cannot go from one side of a surface to another without crossing over an edge of the surface, this shows that the twisted band has only one side. It is called a Möbius band, after the German mathematician who discovered it.

In addition to having only one side, the Möbius band has only one edge. You can check this by coloring in the edge of the band with a pencil. If you color an edge of an ordinary (i.e., cylindrical) band, one edge will remain uncolored. But if you do it for a Möbius band, you will discover that there is no uncolored edge. The entire single edge is colored.

As the example of the Möbius band suggests, sidedness is bound up with the existence of edges. For the most part, mathematicians concentrate on surfaces that have no edges—what they call closed surfaces. The reason is partly that an edge is not really part of a surface. Furthermore, the more interesting topological properties have to do with the surface's internal structure—how it twists and turns. In fact, for each surface that has one or more edges, there is generally a closed surface that has more or less the same characteristics. For example, a sphere has similar characteristics to a finite plane (such as a flat disk), and when we prove a topological result about a sphere, it generally has an immediate consequence for a flat plane (and vice versa). (Intuitively, this is because we can take a perfectly stretchable flat sheet and bring the edge together to form a closed bag—topologically a sphere.)

The closed surface that corresponds to a Möbius band is called a Klein bottle, named after its German discoverer, Felix Klein. The Klein bottle does not have a rim, and has neither

Figure 5.8. A Möbius band, a single-sided, single-edged surface, formed by taking a strip of paper or tape and giving it a half-twist before joining together the two ends. *Source:* Devlin: *Mathematics: The New Golden Age*, Columbia University Press, Figure 50(b), page 228.

an inside nor an outside. (Or to put it another way, the inside and the outside are the same.) Theoretically, you can construct a Klein bottle by taking two Möbius bands and gluing them together along their single edges. I say theoretically because you cannot carry out the gluing in ordinary three-dimensional space. The Klein bottle exists (as a mathematical object) only in four-dimensional space. The best you can do in our three-dimensional world is to allow the surface to pass through itself, in which case you get the object illustrated in Figure 5.9.

Many mathematicians, myself included, have such a self-intersecting Klein bottle made of glass in their offices as an ornament. In four dimensions, the bottle would not have to pass through itself. To the person in the street, an object that exists only in four-dimensional space doesn't really exist, of course, but this trivial objection does not deter the mathematician. After all, everyone "knows" that negative numbers do not have square roots, but that did not prevent mathematicians from developing the complex numbers and, moreover, using them in practical applications. Much of the immense power of mathematics comes

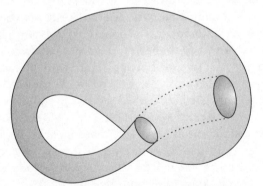

Figure 5.9. Artist's rendering of a Klein bottle, a closed surface that does not divide the surrounding space into an inside and an outside. In three dimensions the Klein bottle can be constructed only if the surface is allowed to pass through itself. In four-dimensional space it can be constructed without self-intersection. *Source:* Devlin: *Mathematics: The New Golden Age*, Columbia University Press, Figure 44, page 184.

from the fact that we can use it to investigate objects that are beyond our ordinary conception as living creatures in a three-dimensional world.

For instance, we can investigate the properties of networks drawn on a Klein bottle. When we do, we discover that the Euler characteristic of a Klein bottle (i.e., the value of the expression $V - E = F$) is 0, the same as for the torus. Aha! Does this mean that the Klein bottle is topologically equivalent to the torus? No. The Euler characteristic cannot distinguish between a Klein bottle and a torus, but the chromatic number can. The chromatic number of a Klein bottle is 6, whereas for the torus it's 7.

The topological property of a Klein bottle that corresponds to the one-sided nature of its surface is a strange concept called nonorientability. What this means is that on the surface of a Klein bottle, you cannot distinguish left-handedness from right-handedness or clockwise rotation from counterclockwise. If you were to draw a small left hand on the surface of a Klein bottle, and then slide the drawing around the surface sufficiently far (far enough so that, if the Klein bottle were in three-dimensional space, the hand would pass right through the self-intersecting neck), then when it returned to the starting point, you would find that it had miraculously turned into a right hand. This experiment is easier to do with a Möbius band. Draw a small left hand on the surface, and then repeatedly make adjacent copies until you get back to your starting point. You will find that by the time you get back to your starting point, the left hand will have become a right hand. Alternatively, if you draw a small circle on the surface of a Klein bottle or a Möbius band, with an arrowhead indicating clockwise rotation, and if you slide or copy the figure around the entire surface until you come back to your starting point, you will find that the arrowhead now points in the counterclockwise direction. See Figure 5.10.

A surface on which you cannot turn left-handed into right-handed or clockwise into counterclockwise by sliding figures around the surface is called orientable. For example, a sphere (or a plane) is orientable and so are a torus and a double torus. A surface, such as a Klein bottle (or a Möbius band), where such

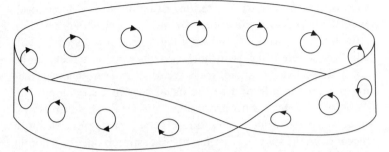

Figure 5.10. Nonorientability of a Möbius band. As the arrow is pushed all the way around the band, its orientation changes. *Source:* Devlin: *Mathematics: The New Golden Age*, Columbia University Press, Figure 52, page 230.

changes can be made is called nonorientable. Orientability (or nonorientability) is a topological invariant.

An Urge to Classify

One of the early successes of topology was to show that just two topological invariants, the Euler characteristic and orientability, are all you need to be able to distinguish any two closed surfaces. That is to say, if two surfaces have the same Euler characteristic and are either both orientable or both nonorientable, then they are in fact the same—even if you can't for the life of you see how to continuously deform one into the other. This result is called the classification theorem for surfaces, since it says that you can classify all surfaces (topologically) by means of just these two attributes.

Loosely speaking, the classification theorem for surfaces is proved by taking a sphere as the basic surface and measuring the degree to which any given surface differs from a sphere—what would you have to do to a sphere to turn it into that surface. This corresponds to our ordinary intuition that a sphere is the simplest, most basic, and, some might say, the most aesthetically perfect closed surface.

I should point out that in this case, the operations to be performed on a sphere to turn it into some other surface go beyond the normal topological operations of continuous deformation. Indeed, if you change a sphere by means of twisting, bending, stretching, or shrinking, the resulting object, topologically, will still be a sphere. To classify surfaces by seeing how they can be constructed from a sphere, you have to allow cutting and stitching together in addition to the usual twisting, stretching, etc. Topologists refer to this process as "surgery." The term is apt, since a typical surgical operation involves cutting one or more pieces from the sphere, twisting, turning, stretching, or shrinking each of those pieces, and then sewing those pieces back into the sphere again.

The classification theorem tells us that any orientable surface is topologically equivalent to a sphere with a certain number of "handles" sewn onto it. You get a handle by cutting two holes into the sphere and joining them together by means of a tube, as in the left side of Figure 5.11. Any nonorientable surface is equivalent to a sphere with a certain number of "crosscaps" sewn in. You get a crosscap by cutting a hole in the sphere and sewing a Möbius band to the boundary of the hole, as shown in the right side of Figure 5.11. As with the Klein bottle, in ordinary three-dimensional space you can't do this without the Möbius band passing through itself; you need four dimensions to do it properly.

In the early years of the twentieth century, Poincaré and other mathematicians set out to classify higher-dimensional analogues of surfaces—which they called "manifolds." Not surprisingly, they tried an approach similar to the one that had worked for two-dimensional surfaces. They sought to classify all three-dimensional manifolds (called "3-manifolds" for short) by taking a three-dimensional analogue of a sphere (called a "3-sphere") as basic and measuring the degree to which any 3-manifold differs from that 3-sphere.

We need to be careful here. A regular surface such as a sphere or a torus is a two-dimensional object. The figure the surface encloses is three-dimensional, of course, but the surface itself is two-dimensional. Apart from a plane, any surface can be constructed only in a space of three or more dimensions. Thus, any

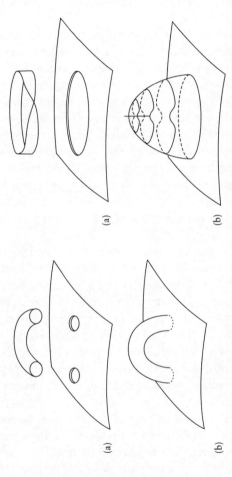

Figure 5.11. Handles and crosscaps. To create a handle (left figure) on a surface, cut two small holes in the surface and join them by a cylindrical tube. To create a crosscap (right figure), cut one hole and sew a Möbius to the boundary of the hole. Since the Möbius band has only one edge, this is conceptually possible. However, in three dimensions it can be done only if the band is allowed to intersect itself. A fundamental theorem of topology—the Classification Theorem for surfaces—says that any smooth, closed surface is topologically equivalent to a sphere with a fixed number of handles or crosscaps. *Source:* Devlin: *Mathematics: The New Golden Age,* Columbia University Press, Figure 59, page 249, and Figure 60, page 250.

closed surface requires three or more dimensions. For instance, it takes three dimensions to construct a sphere or a torus, four dimensions to construct a Klein bottle. Yet a sphere, a torus, or a Klein bottle is a two-dimensional object—a surface that has no thickness and can, in principle, be constructed from a flat, perfectly elastic sheet.

But just as a sphere can be regarded as a two-dimensional analogue (in three-dimensional space) of a circle (which is a one-dimensional object—a curved line—in two-dimensional space), so too we can imagine a three-dimensional analogue (in four-dimensional space) of a sphere. Well, actually, we can't imagine it. But we can write down equations that determine such an object, and study "it" mathematically. Indeed, physicists routinely study such imaginary objects, and use the results to help understand the universe we live in. The 3-manifolds, i.e., the three-dimensional analogues of surfaces (which exist in spaces of four or more dimensions), are sometimes called hypersurfaces, with the three-dimensional analogue of a sphere being called a hypersphere.

There is no mathematical reason to stop at three dimensions. You can write down equations that determine manifolds of 3, 4, 5, 6, or any number of dimensions. Once again, these considerations turn out to be more than idle speculation. The mathematical theories of matter that physicists are currently working on view the universe we live in as having 11 dimensions. According to these theories, we are directly aware of three of those dimensions, and the others manifest themselves as various physical features such as electromagnetic radiation and the forces that hold atoms together.

Poincaré attempted to classify manifolds of three and more dimensions by taking a "sphere" of the respective dimension as a base figure and then applying surgery. A natural first step in this endeavor was to look for a simple topological property that tells you when a given (two-dimensional) surface is topologically equivalent to a sphere. (Remember, we are doing topology here. Even in the simple case of regular two-dimensional surfaces, a surface might appear extremely complicated and yet turn out to be continuously deformable to a sphere.)

In the case of two-dimensional surfaces, there is such a property. Suppose you were to take a pencil and draw a simple closed loop on the surface of a sphere. Now imagine the loop shrinking in size, sliding over the surface as it does so. Is there a limit to how small the loop can shrink? Obviously not. You can shrink the loop until it becomes indistinguishable from a point. Mathematically, you can shrink it until it actually becomes a point.

The same thing is not necessarily true if you start with a loop drawn on a torus. You can draw loops on a torus that cannot be shrunk down to a point. No loop that goes right around the ring of the torus can be shrunk down indefinitely, nor can any loop that encircles the torus like a belt.

The shrinkability to a point of any loop drawn in a surface is a topological property of the surface that is unique to spheres. That is to say, if you have a surface on which every loop (the "every" is important here) can be shrunk down to a point without leaving the surface, then that surface is topologically equivalent to a sphere.

Is the same true for a three-dimensional hypersphere? This is the question Poincaré asked in the early 1900s, hoping that a speedy positive answer would be the first step on the road to a classification theorem for three-dimensional hypersurfaces. He developed a systematic method—called homotopy theory— for studying (using methods of algebra) what happens to loops when they are moved around a manifold and deformed.

Actually, that's not quite what happened. At first, Poincaré tacitly *assumed* that the loop-shrinking property for 3-manifolds did characterize the 3-sphere. After a while, however, he realized that his assumption might not be valid, and in 1904 he put his doubts into print, writing (in French), "Consider a compact three-dimensional manifold V without boundary. Is it possible that the fundamental group of V could be trivial, even though V is not homeomorphic to the three-dimensional sphere?" Stripping away the technical terms, what Poincaré asked was, "Is it possible that a 3-manifold can have the loop-shrinking property and not be equivalent to a 3-sphere?" That was the birth of the Poincaré conjecture.

As it turned out, his question did not get a speedy answer. Nor, indeed, a slow answer, despite the best efforts of a number of leading topologists. As a result, finding a proof (or a disproof) of the Poincaré conjecture rose to become one of the most sought-after prizes in mathematics.

Progress of any kind was a long time coming. In 1960, the American mathematician Stephen Smale proved that the Poincaré conjecture is true for manifolds of all dimensions from five upward. Thus, if a five or more dimensional manifold has the property that any closed loop drawn on it can be shrunk down to a point, then that manifold is topologically equivalent to a hypersphere of the appropriate dimension.

Unfortunately, Smale used methods that did not work for manifolds of dimension three or four, so the original Poincaré conjecture remained unresolved. Then, in 1981, another American, Michael Freedman, found a way to prove the conjecture for four-dimensional manifolds. (Freedman's work turned out to be extremely useful to physicists studying the nature of matter.)

And there the issue rests. The Poincaré conjecture has been shown to be true for every dimension except three—the very dimension for which Poincaré originally raised the question. For their achievements, both Smale and Freedman were awarded Fields Medals—generally regarded as the mathematicians' equivalent of a Nobel Prize. A similar honor will doubtless be granted to the first person to prove the one remaining case of the Poincaré conjecture. (Provided that individual is younger than the cut-off age of forty for an award of a Fields Medal. The medals were established to encourage young mathematicians to tackle the major problems of the discipline.) Whatever the age of the solver, however, he or she will now receive not only the accolades of the entire mathematical community, but also a $1 million Millennium Prize.

Should the would-be solver look for a proof of the conjecture or seek a counterexample—a three-dimensional hypersurface that has the loop-shrinking property but is not topologically equivalent to a three-dimensional hypersphere? Not surprisingly, given that it has been proved for all other dimensions, the smart money goes on the Poincaré conjecture being true. It could well be, however, that the proof will stretch into hundreds of pages. The last time someone produced a purported proof that the mathematical community took seriously, some twenty years ago, it took several months of close examination before it was finally agreed that the proof had a fatal flaw.

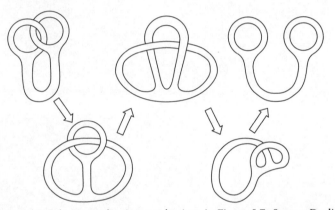

Figure 5.12. Solution to the ring puzzle given in Figure 5.7. *Source:* Devlin: *Mathematics: The New Golden Age*, Columbia University Press, Figure 62, page 261.

SIX

Knowing When the Equation Can't Be Solved

The Birch and Swinnerton-Dyer Conjecture

In the early 1960s, computers were still in an early stage of development, and there were only a few in existence, mostly located at major universities. As faculty members at Cambridge University in England, the British mathematicians Brian Birch and Peter Swinnerton-Dyer had access to what was then one of the world's most powerful computing machines, the Cambridge EDSAC. They set out to use the computer to try to gather some data about the possible solutions to certain kinds of polynomial equations. The data they obtained—more precisely, the patterns they discerned—eventually led them to formulate a bold and powerful conjecture that, if true, would have major consequences for our understanding of the whole numbers. Their conjecture now has a $1 million price on its head.

The Birch and Swinnerton-Dyer conjecture involves mathematical objects called elliptic curves. These are not the same as ellipses. (You can see two elliptic curves by glancing forward to Figures 6.3 and 6.4.) The name "elliptic curve" comes from the fact that you encounter them (more precisely, you encounter their equations) when you compute the arc lengths of ellipses.

Since the early 1950s it has become clear to mathematicians that elliptic curves are important and fundamental objects that have connections to many areas of mathematics, including number theory, geometry, cryptography, and the mathematics of data transmission. For instance, when Andrew Wiles proved Fermat's last theorem in 1994, he did it by proving a result about elliptic curves—indeed by establishing a close connection between elliptic curves and another important part of mathematics, the theory of modular forms. (Don't even ask what those are. They cannot be described in simple terms—at least it's beyond my ability to do so.) A proof of the Birch and Swinnerton-Dyer conjecture will have ramifications throughout modern mathematics.

Although the conjecture itself is buried deep in highly advanced mathematics, we can approach it from some very humble beginnings: Pythagoras's theorem and the formula for calculating the area of a triangle.

Half the Base Times the Height

A classical problem, dating back to the ancient Greeks, goes as follows. Suppose you are given a positive whole number d. Is there is a right triangle whose sides are rational numbers (i.e., whole numbers or fractions) for which the area is exactly d? For example, in the case $d = 6$, the answer is yes. The famous Pythagorean right triangle with sides 3, 4, and 5, shown in Figure 6.1, has area

$$A = \tfrac{1}{2} \times \text{base} \times \text{height} = \tfrac{1}{2} \times 4 \times 3 = 6$$

In the case $d = 5$, there is no right triangle with whole-number sides that has area 5, but the right triangle with sides $\frac{3}{2}$, $\frac{20}{3}$, $\frac{41}{6}$, shown in Figure 6.2, has area 5.

It is a fairly straightforward piece of algebraic reasoning to show that there is a right triangle with rational sides having an area d if and only if the equation

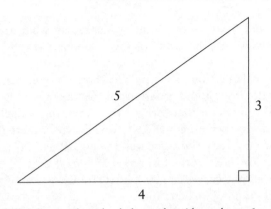

Figure 6.1. A right triangle with whole-number sides and area 6.

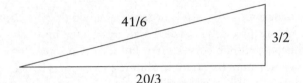

Figure 6.2. A right triangle with rational sides and area 5.

$$y^2 = x^3 - d^2 x$$

has rational solutions for x and y with $y \neq 0$.[1]

Equations of the general form

$$y^2 = x^3 + ax + b$$

---

1. For those who are interested, here is the main idea of that argument. First, if there is a right triangle with rational sides a, b, c and area d, then $x = \frac{1}{2}a(a-c)$, $y = \frac{1}{2}a^2(c-a)$ solves the equation. Conversely, if x and y are rational numbers such that $y^2 = x^3 - d^2 x$, with $y \neq 0$, then the triangle with sides

$$\left| \frac{x^2 - d^2}{y} \right|, \quad \left| \frac{2xd}{y} \right|, \quad \left| \frac{x^2 + d^2}{y} \right|$$

is right-angled and has area d.

where a and b are whole numbers determine what are called elliptic curves, i.e., the graph of such an equation is an elliptic curve.[2]

A natural question to ask is, Why there is no term involving x^2? Why don't we allow equations of the kind $y^2 = x^3 + ax^2 + bx + c$? The answer is that with a fairly simple bit of algebra you can transform such an equation to one in which there is no x^2 term. Thus, to study elliptic curves, the only equations you need to look at are those of the form $y^2 = x^3 + ax + b$.

Some elliptic curves can look a bit strange at first. If you try to draw a graph of an equation

$$y^2 = x^3 + ax + b$$

then whenever $x^3 + ax + b$ is negative, you won't get an answer for y. (More precisely, y will be an imaginary number.) As a consequence, elliptic curves often fall into two separate pieces, as shown in Figure 6.3. (For an elliptic curve that is in one piece, see Figure 6.4. Whether an elliptic curve is in one piece or two depends on whether the cubic expression on the right of the equation has one real root or three.)

What about the triangle area equation $y^2 = x^3 - d^2x$, where d is a whole number? As we noted, this equation has rational-number solutions for x and y precisely when d is the area of a right triangle with rational sides. For this equation the discriminant is $\Delta = -16(-4d^2) = 64d^2$, which is nonzero, so the graph of the equation is an elliptic curve. (Put $a = -d$ and $b = 0$ in the formula for the discriminant.) Thus, the ancient Greek problem of finding whole numbers d that are areas of right triangles with rational sides is equivalent to the problem of finding rational points (i.e., points whose coefficients are rational numbers) on certain elliptic curves. That is the problem Birch and Swinnerton-Dyer set out to investigate.

---

2. Strictly speaking, for the graph to be an elliptic curve, the equation should satisfy an additional condition: Its discriminant should be nonzero. The discriminant is the quantity $\Delta = -16(4a^3 + 27b^2)$.

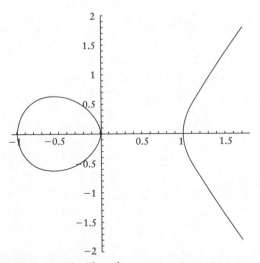

Figure 6.3. The elliptic curve $y^2 = x^3 - x$. Although it splits into two separate pieces, it is a single curve, determined by a single equation.

The two researchers approached the task by trying to find some way of "counting" the number of rational points on elliptic curves. Of course, since they were dealing with collections that could be infinite, the word "counting" has to be taken somewhat metaphorically.

One way to count a possibly infinite collection that sometimes works is to carry out a series of finite subcounts. This is what Birch and Swinnerton-Dyer did. To describe their method, we first have to break off and talk about finite arithmetic.

Counting by the Clock: Finite Arithmetic

We are all familiar with one situation where we count an endless (and hence potentially infinite) collection: the way we count minutes. There is (let's be optimistic) no last minute—time goes on forever. And yet we count minutes using just sixty numbers, 0 through 59. What we do, of course, is keep restarting the count. After we reach 59 minutes, we start again at 0. Another way to say this is that we treat 60 as if it were 0. Mathematicians would

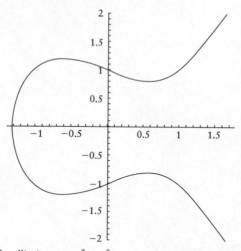

Figure 6.4. The elliptic curve $y^2 = x^3 - x + 1$. An example of an elliptic curve that does not split into two pieces.

say that we count minutes *modulo* 60. (The number 60 is the *modulus* for this counting.) Similarly, we count hours modulo 12 (or modulo 24).

For any positive whole number N, we can count modulo N. The numbers we use for this counting are $0, 1, 2, \ldots, N - 1$. After $N - 1$, we start again at 0.

We can then do arithmetic modulo N. To explain how, let's take the case $N = 7$. For this modulus, the counting numbers are $0, 1, 2, 3, 4, 5, 6$. When we add any two numbers in this range, we discard 7 whenever it arises. For example, in arithmetic modulo 7,

$$2 + 3 = 5, \quad 3 + 4 = 0, \quad 4 + 5 = 2, \quad 6 + 6 = 5$$

Of course, this looks strange, and could be confusing, so mathematicians don't write it this way. Instead, they express the above additions like this:

$$2 + 3 \equiv 5 \quad (\text{mod } 7)$$

$$3 + 4 \equiv 0 \quad (\text{mod } 7)$$

$$4 + 5 \equiv 2 \quad (\text{mod } 7)$$

$$6 + 6 \equiv 5 \quad (\text{mod } 7)$$

They call such expressions *congruences*. The first of the above four congruences is read "two plus three is congruent to five modulo seven." Notice that there is nothing special about 7 here. You could do the same thing with any other number.

Multiplication modulo 7 is defined similarly: You multiply the two numbers in the usual way and discard any multiples of 7. Thus

$$2 \times 3 \equiv 6 \quad (\text{mod } 7)$$

$$3 \times 4 \equiv 5 \quad (\text{mod } 7)$$

$$4 \times 5 \equiv 6 \quad (\text{mod } 7)$$

$$6 \times 6 \equiv 1 \quad (\text{mod } 7)$$

Similarly for any other modulus.

Since subtraction is the opposite of addition, you can always subtract in finite arithmetic. For example,

$$5 - 3 \equiv 2 \quad (\text{mod } 7)$$

$$3 - 5 \equiv 5 \quad (\text{mod } 7)$$

$$4 - 5 \equiv 6 \quad (\text{mod } 7)$$

$$1 - 6 \equiv 2 \quad (\text{mod } 7)$$

(To check these, simply add to both sides—modulo 7—the second term on the left.) The same is true for any modulus.

How about division? In the case where the modulus is 7, you can always divide. For example,

$$5 \div 3 \equiv 4 \quad (\text{mod } 7)$$

$$3 \div 5 \equiv 2 \quad (\text{mod } 7)$$

$$4 \div 5 \equiv 5 \quad (\text{mod } 7)$$

$$1 \div 6 \equiv 6 \quad (\text{mod } 7)$$

(To check these, simply multiply—modulo 7—both sides by the second term on the left.) In fact, division works whenever the modulus is a prime number. But for a composite modulus, it is not always possible to divide one number by another in finite arithmetic. (Of course, you can't always divide one whole number by another in ordinary arithmetic if you want the answer to be a whole number as well.)

Thus, modular arithmetic—as these finite versions of arithmetic are sometimes called—is just like regular arithmetic. If the modulus is prime, then the corresponding modular arithmetic even has the additional property that you can divide any number by another (and get a whole-number answer).

Modular arithmetic has proved useful on a number of occasions. One of those was in providing a way for Birch and Swinnerton-Dyer to count the rational points on an elliptic curve.

How to Count an Infinite Set

In order to count rational points on an elliptic curve, Birch and Swinnerton-Dyer decided to carry out analogous counts modulo p for various primes p. That is to say, instead of trying to count the possibly infinite number of rational solutions to an equation

$$y^2 = x^3 + ax + b$$

they took different prime numbers p and counted the number of pairs (x, y) of integers modulo p such that

$$y^2 \equiv x^3 + ax + b \quad (\text{mod } p)$$

For any given prime p, this counting is finite, of course, and hence can actually be carried out. Let N_p be the number of solutions modulo p; i.e., N_p is the number of pairs (x, y) of integers modulo p such that

$$y^2 \equiv x^3 + ax + b \pmod{p}$$

For example, suppose we take the elliptic curve $y^2 = x^3 - x$ and the prime $p = 5$. Then, by testing the congruence

$$y^2 \equiv x^3 - x \pmod{5}$$

with all possible pairs of values (x, y) for $x = 0, 1, 2, 3, 4$ and $y = 0, 1, 2, 3, 4$, we find that the solutions are $(0, 0), (1, 0), (4, 0),$ $(2, 1), (3, 2), (3, 3), (2, 4)$. There are 7 of these. Hence, for this equation, $N_5 = 7$.

The idea behind looking at the finite, mod p, versions of the counting problem is this: If (u, v) is a whole-number solution to the equation

$$y^2 = x^3 + ax + b$$

then $(u \bmod p, v \bmod p)$ solves the congruence

$$y^2 \equiv x^3 + ax + b \pmod{p}$$

where $u \bmod p$ is the remainder on dividing u by p, etc. More generally, because division modulo a prime modulus always leads to a whole-number answer, any rational solution to the original equation gives rise to a whole-number solution to the corresponding congruence. Thus, if there is a rational point on the original elliptic curve, then for every prime number p, the corresponding mod p congruence has a solution. If in fact there are infinitely many rational points on the elliptic curve, we can expect that for many primes p the congruence has many solutions. (This last observation gains significance because, as we shall see presently, in the case of an elliptic curve that arises from the triangle area problem, if there is one rational point on the curve, then there are infinitely many.)

There is no obvious reason why the converse would be true: that the existence of many solutions to the mod p congruences for lots of primes p implies that the original equation definitely has a rational solution, let alone infinitely many. But it would surely seem a likely possibility—or so Birch and Swinnerton-Dyer assumed. More precisely, they based their conjecture on the assumption that the existence of *lots* of solutions to the congruences for *lots* of primes would imply that the original equation does indeed have infinitely many rational solutions.

The question then was, how do you find out whether there are lots of solutions to lots of those congruences?

Now, if you have reached this point, you will have a general idea of what the Birch and Swinnerton-Dyer conjecture is about, and how it relates to a classic geometry problem about right triangles. You should feel pretty good about having gotten this far. Unfortunately, the going is going to get quite a bit harder from this point on. Don't feel bad if you find yourself getting lost. Most readers will. Like many parts of modern advanced mathematics, the level of abstraction is simply too great for the nonexpert to make much headway. Although I have been a professional mathematician for over thirty years, number theory is not my area of expertise, and it took me considerable effort, spread over several weeks, aided by discussions with experts that I knew, before I understood the problem sufficiently to write this chapter. I would not even attempt to try to solve it.

If you still want to continue, let's pick up the thread again. (When you feel you cannot proceed any further, simply give up and start the next chapter—where, I have to be honest, you are likely to make even less progress than you have here.)

To determine whether there are lots of solutions to lots of those mod p congruences, Birch and Swinnerton-Dyer computed the "density functions"

$$\prod_{p \leq M, \ p \ \text{prime}} \frac{p}{N_p}$$

(where N_p is as above) for larger and larger values of M.

[If you are not familiar with the $\prod$ notation used above, or with the $\sum$ notation to be used momentarily, see the appendix to this chapter for an explanation.]

The next step was to examine the data they got—primarily graphs of the values of $\prod_{p \leq M} \frac{p}{N_p}$ for increasing values of M—and try to find some formula that described the data. The obvious formula to look at first was the infinite product

$$\prod_p \frac{p}{N_p}$$

taken over *all* primes. If this infinite product were guaranteed to give a finite answer, the values of $\prod_{p \leq M} \frac{p}{N_p}$ that Birch and Swinnerton-Dyer computed for larger and larger values of M would have provided a sequence of approximations to that infinite product, and they could have used the infinite product to analyze their computational data. Unfortunately, the infinite product cannot be guaranteed to give a finite answer. Nevertheless, the strategy of looking for a formula that captured the data was a good one, and it turned out that a related formula does work. Since that other infinite product is more complicated than the one above, however, what I'll do is outline the way the analysis would have gone had the infinite product above given an answer, and then describe the changes that Birch and Swinnerton-Dyer made in order to get an argument that worked.

If the original elliptic curve has infinitely many rational points, then for many primes p the mod p congruence should have a large number of solutions, which means that for infinitely many primes the ratio $\frac{p}{N_p}$ should be (much) less than 1, and hence the infinite product should work out to be 0. The conjecture Birch and Swinnerton-Dyer made was that this argument would work the other way round: If we calculate $\prod_p \frac{p}{N_p}$ and find that it is zero, then maybe that will tell us that the elliptic equation does in fact have infinitely many rational points. In other words, perhaps the elliptic curve will have an infinite number of rational points if and only if

$$\prod_p \frac{p}{N_p} = 0$$

But how do you work out this infinite product? Well, we've met an infinite product of fractions taken over all the primes before, in Chapter 1. Euler showed that for any real number $s > 1$, the infinite product

$$\prod_{p \text{ prime}} \frac{1}{1 - (1/p^s)}$$

is equal to the infinite sum

$$\zeta(s) = \sum_{n=1}^{\infty} \frac{1}{n^s}$$

Riemann (and others) then went on to show that the function $\zeta(s)$ could be extended to give an answer for any complex number s, and that the extended function could be studied using methods of calculus. Dirichlet showed that the same kind of process would work for a more general class of "zeta functions," called L-functions. (See the appendix to Chapter 1.)

Suppose you could do the same kind of thing for the Euler-like infinite product

$$\prod_p \frac{p}{N_p}$$

That is, suppose you could show that there is a function $L(E, s)$ that gives an answer for any complex number s and that can be studied using methods of calculus, such that

$$L(E, 1) = \prod_p \frac{p}{N_p}$$

(The E is included in the notation for L because the numbers N_p depend on E.) Then, by calculating $L(E, 1)$, you could get some information about the number of rational points on the elliptic curve. In fact, said Birch and Swinnerton-Dyer, you might get everything you wanted. Based on the evidence from their computer runs, they suggested that the elliptic curve will have an infinite number of rational points if and only if $L(E, 1) = 0$.

This, in essence, is the Birch and Swinnerton-Dyer conjecture. But note that modifier "in essence." If you try to do it exactly as I've described it, it won't work. To have any hope of carrying out a Dirichlet-like argument to get an "L-function," you need to take a slightly more complicated product than $\prod_p \frac{p}{N_p}$. (For one thing, as I mentioned, this simple infinite product does not give a finite answer.) Then there's the question of whether there really is a suitable L-function (that gives an answer for all complex numbers s). That turns out to be a special case of the Tanayama–Shimura conjecture, which was not resolved until 1994, when Andrew Wiles and Richard Taylor proved it en route to the solution of Fermat's last theorem.[3] Prior to 1994, it was not even certain that the Birch and Swinnerton-Dyer conjecture really made sense. No one knew whether there was a function $L(E, s)$; more precisely, no one knew whether there was such a function that gave an answer for all numbers s, in particular for the key value $s = 1$. Now that we know such a function does indeed exist, the big question is whether the conjecture Birch and Swinnerton-Dyer made about it is true, namely, that there are infinitely many rational points on E if and only if $L(E, 1) = 0$.

In the remainder of this chapter I'll provide a bit more detail—and a bit more precision—about the argument I've just outlined.

Why Elliptic Curves Are Important: The Group Structure

Much of the interest in elliptic curves, and the reason why they arise all over the place in modern mathematics, is bound up with

---

3. Strictly speaking, Wiles and Taylor proved only part of the conjecture. The missing pieces were supplied in 1999 by Christophe Breuil, Brian Conrad, Fred Diamond, and Richard Taylor.

the fact first observed by Henri Poincaré in 1901 that with each elliptic curve there is associated a particular group. (We met the group concept in Chapter 2. You might need to look back over the appendix to that chapter before proceeding further.)

The objects that make up the group are the points on the elliptic curve whose coefficients are rational numbers. Mathematicians generally use the symbol $\mathbb{Q}$, a stylized capital Q, to denote the set of rational numbers. Thus, the group is made up of points (x, y) on the curve where both x and y are in the set $\mathbb{Q}$. Mathematicians express this in the following more formal way: Given an elliptic curve E, let $E(\mathbb{Q})$ be the set of all rational points on E (i.e., all points on E whose coefficients are in $\mathbb{Q}$). Actually, that's not quite right. For technical reasons, you have to throw one additional point into the set $E(\mathbb{Q})$, namely the idealized "point at infinity", ∞, which lies on all vertical lines. (For example, for the curve $y^2 = x^3 - x$, shown in Figure 6.3, $E(\mathbb{Q}) = \{(0,0), (1,0), (-1,0), \infty\}$.)

To make $E(\mathbb{Q})$ into a group, you have to say what it means to add together two members of this set. This is shown in Figure 6.5.

With addition defined as in Figure 6.5, $E(\mathbb{Q})$ is a group. In fact, it is a commutative (or abelian) group. In 1922, the English mathematician Lewis Mordell proved that the group $E(\mathbb{Q})$ is finitely generated. That is, even if $E(\mathbb{Q})$ has infinitely many members, there are finitely many points in $E(\mathbb{Q})$ such that every member of $E(\mathbb{Q})$ can be reached by a finite sequence of additions starting from those finitely many points. In geometric terms, there is a finite collection of rational points on the curve such that every rational point on the curve can be reached by means of a finite sequence of chord/tangent-line steps—as described in Figure 6.5—starting from that initial finite collection.

For any group, if we start with a single point A and generate the sequence of points $A, A + A, A + A + A, \ldots$, then one of two things will happen. Either this sequence of points will eventually cycle back onto itself, or else it will continue forever. In the former case, the sequence of points produces a finite subgroup of the group; in the latter case, it produces a copy of the

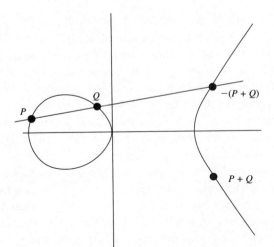

Figure 6.5. Group addition for an elliptic curve. Given points P and Q in $E(\mathbb{Q})$, draw the line through P and Q. It will intersect the curve in at most one other point. If this line does not meet the curve again, we say that the chord "meets the curve at ∞," the point at infinity. With this provision, given any two points P and Q, the chord PQ, when extended, meets the curve in exactly one other point (possibly ∞). That third point is taken to be $-(P + Q)$. More precisely, the group sum $P + Q$ is taken to be the point on the curve directly opposite the third point, when reflected in the x-axis. (The reflection of ∞ is again ∞.)

integers, generally denoted by the symbol $\mathbb{Z}$, a stylized letter Z. (After the German word for numbers, *Zahlen*.) In the case of the group $E(\mathbb{Q})$, because it is finitely generated, if we look at what happens when we carry out the above iteration to each of the generators, we see that the group splits up into a finite number of finite subgroups and a finite number of copies of $\mathbb{Z}$.

In the more formal language of mathematics,

$$E(\mathbb{Q}) \cong \mathbb{Z}^r \times E(\mathbb{Q})_f$$

where $\mathbb{Z}$ is the (infinite) group of integers under addition, r is some nonnegative integer, and $E(\mathbb{Q})_f$ is a finite group (the subgroup of $E(\mathbb{Q})$ of all members of finite order). The number r is called the rank of the curve E. As we shall see, the number r provides a way to measure the size of the (possibly infinite) set of rational points on the curve; the larger r is, the more rational points we expect to see. Thus, the rank of an elliptic curve is an important parameter. However, although there has been considerable progress in the theory of elliptic curves in recent years, the rank remains mysterious. Even such basic questions as to how to compute the rank or whether the rank can be arbitrarily large remain unsettled.

During the 1930s, Trygve Nagell and Elisabeth Lutz independently proved that if a point (x, y) in $E(\mathbb{Q})$ has finite order (i.e., is a member of $E(\mathbb{Q})_f$), then x and y must both be whole numbers, and either $y = 0$ or else y^2 evenly divides Δ, the discriminant of the equation E. More recently, in 1977, Barry Mazur showed that $E(\mathbb{Q})_f$ has to be one of the 11 groups $\mathbb{Z}/n\mathbb{Z}$, $n = 1, 2, \ldots, 10, 12$ or else one of the 4 groups $(\mathbb{Z}/m\mathbb{Z}) \times (\mathbb{Z}/2\mathbb{Z})$, $m = 1, 2, 3, 4$. In the case where E is $y^2 = x^3 - d^2x$ (i.e., a curve that comes from the triangle area problem), $E(\mathbb{Q})_f = (\mathbb{Z}/2\mathbb{Z}) \times (\mathbb{Z}/2\mathbb{Z})$. See Figure 6.6.

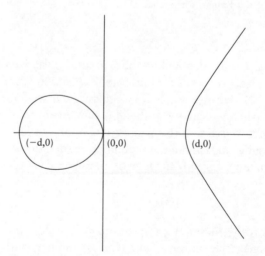

Figure 6.6. The elliptic curve $y^2 = x^3 - d^2x$.

Counting Rational Points on Elliptic Curves

We have already observed that there is a rational-sided right triangle with area d if and only if the elliptic curve $y^2 = x^3 - d^2x$ has a rational point with $y \neq 0$. But the rational points with $y \neq 0$ are exactly the points of infinite order. Hence, there is a rational-sided right triangle with area d if and only if the elliptic curve $y^2 = x^3 - d^2x$ has infinitely many rational points. (A startling consequence of this is that if there is a rational-sided right triangle with area d, then there are infinitely many of them.)

Thus, as we observed earlier, the original question about right triangles can be solved by "counting" the number of rational points on certain elliptic curves. The idea, you will remember, is to count solutions modulo prime numbers p.

For any prime number p, let N_p be the number of pairs (x, y) of integers modulo p such that

$$y^2 \equiv x^3 + ax + b \pmod{p}$$

plus 1. (Notice that this is 1 more than the N_p we introduced earlier. The 1 is added to account for the point ∞.) Mathematicians will realize that this new N_p is the order of the group $E(\mathbb{Z}/p\mathbb{Z})$.

In the case of our earlier example of the elliptic curve $y^2 = x^3 - x$ and the prime $p = 5$, we found that there are 7 solutions, namely $(0, 0)$, $(1, 0)$, $(4, 0)$, $(2, 1)$, $(3, 2)$, $(3, 3)$, $(2, 4)$, and hence, for this equation, $N_5 = 7 + 1 = 8$.

The idea behind looking at the finite, mod p, versions of the counting problem, remember, is that if an elliptic curve $y^2 = x^3 + ax + b$ has infinitely many rational solutions, then for different primes p, the congruence

$$y^2 \equiv x^3 + ax + b \pmod{p}$$

should tend, on average, to have a large number of solutions.

The idea is to test this by computing some form of density function. Earlier work by Helmut Hasse and André Weil had led to the formulation of the following function that seemed to be what Birch and Swinnerton-Dyer needed:

$$L(E, s) = \prod_p \left(1 - \frac{1 + p - N_p}{p^s} + \frac{p}{p^{2s}} \right)^{-1}$$

This is generally called the Hasse–Weil L-function. It would take us too far from our path to explain where the particular terms in this expression come from. Notice, however, that if you simply substitute $s = 1$ in the above expression and simplify the algebra, you get the infinite product

$$\prod_p \left(\frac{N_p}{p} \right)^{-1} = \prod_p \frac{p}{N_p}$$

Strictly speaking, this algebra is meaningless, because when you set $s = 1$ the resulting infinite product does not give an answer. Nevertheless, it does indicate how the Hasse–Weil function relates to our earlier intuitive discussion of the Birch and Swinnerton-Dyer method.

Let me try to give some idea of why the Hasse–Weil formula met the requirements of Birch and Swinnerton-Dyer. Hasse had proved that the numbers N_p are for the most part roughly equal to $p + 1$, with a maximum variation around this value of $2\sqrt{p}$. Putting this another way, if we let a_p be the "variation" amount

$$a_p = (p + 1) - N_p$$

between N_p and $p + 1$, then $|a_p| < 2\sqrt{p}$. In terms of these a_p terms, the Hasse–Weil L-function can be written like this:

$$L(E, s) = \prod_p \left(1 - \frac{a_p}{p^s} + \frac{p}{p^{2s}} \right)^{-1}$$

Hasse's inequality $|a_p| < 2\sqrt{p}$ then implies that $L(E, s)$ gives an answer whenever the real part of s exceeds $\frac{3}{2}$.

Intuitively, the terms a_p/p^s in the above product can be thought of as "correction" terms to compensate for the variation of N_p from $p + 1$. If the a_p are positive and negative in roughly equal numbers, the infinite product $L(E, s)$ might turn out to be nonzero. However, if there is a bias toward the negative—that is, if N_p has a tendency to be bigger than $p + 1$—then $L(E, s)$ could work out to be zero.

Hasse had conjectured that, like the Riemann zeta function, $L(E, s)$ can be extended to a function that gives an answer for any complex number s and to which the methods of calculus can be applied. This bold conjecture was a consequence of the Tanayama–Shimura Conjecture, which was finally resolved in 1999 after Wiles and Taylor proved a special case en route to Fermat's last theorem.

The Birch and Swinnerton-Dyer Conjecture says that $E(\mathbb{Q})$ is infinite if and only if $L(E, 1) = 0$.

In fact, this is not exactly the form in which Birch and Swinnerton-Dyer first formulated their conjecture. The statement they made was somewhat stronger. Suppose, they said, that Hasse's conjecture turned out to be true, and $L(E, s)$ could be extended to a function that gives an answer for any complex number s and to which the methods of calculus can be applied. This would imply, in particular, that $L(E, s)$ could be expressed using what are known as Taylor polynomials. For instance, the values of $L(E, s)$ around the point $s = 1$ (the point of principal interest to the two researchers) could be given by an infinite polynomial of the form

$$L(E, s) = c_0 + c_1(s - 1) + c_2(s - 1)^2 + c_3(s - 1)^3 + \cdots$$

In terms of this polynomial, the conjecture stated above can be reformulated as: $E(\mathbb{Q})$ is infinite if and only if $c_0 = 0$. Birch and Swinnerton-Dyer made the following, much stronger claim: $E(\mathbb{Q})$ is infinite if and only if $c_r \neq 0$ but every coefficient c_n is zero for $n = 0, \ldots, r - 1$, where r is the rank of E. In other words, $E(\mathbb{Q})$ is infinite if and only if the Taylor polynomial for $L(E, s)$ at $s = 1$ has the form

$$L(E, s) = c(s - 1)^r + \text{higher-order terms}$$

where $c \neq 0$ and $r = \mathrm{rank}(E(\mathbb{Q}))$. Intuitively, counting the number of initial terms of a Taylor polynomial that are zero provides a measure of the degree to which the function is zero at the point concerned. Thus, according to Birch and Swinnerton-Dyer's conjecture, the rank of E gives an exact measure of the degree to which $L(E, 1) = 0$.

So now you know.

APPENDIX

NOTATION FOR
INFINITE SUMS AND PRODUCTS

Mathematicians abbreviate a "long" sum such as

$$a_1 + \cdots + a_N$$

(using dotty notation) by

$$\sum_{n=1}^{N} a_n$$

The capital Greek letter Σ is used to stand for "sum." The letter n is an index or counter that runs between the two values indicated, in this case 1 and N.

For example, the sum of the squares of the first 100 natural numbers can be written in this way:

$$\sum_{n=1}^{100} n^2$$

This abbreviates the more explicit (dotty) expression

$$1^2 + 2^2 + 3^2 + \cdots + 99^2 + 100^2$$

The sum of the cubes of the numbers from 100 to 200 can be written like this:

$$\sum_{n=100}^{200} n^3$$

There is an analogous notation for products. The product

$$a_1 \cdot a_2 \cdots a_N$$

is abbreviated by

$$\prod_{n=1}^{N} a_n$$

For example, the product of the squares of all the natural numbers from 50 to 100 can be written either as

$$50^2 \times 51^2 \times 52^2 \times \cdots \times 99^2 \times 100^2$$

or in the abbreviated form

$$\prod_{n=50}^{100} n^2$$

The $\sum$ and $\prod$ notations can be used to express infinite sums and products. For example,

$$\sum_{n=1}^{\infty} a_n = a_1 + a_2 + a_3 + \cdots$$

where the sum here continues forever, and

$$\prod_{n=1}^{\infty} a_n = a_1 \times a_2 \times a_3 \times \cdots$$

where the product continues forever.

Using this notation, Euler's zeta function has the definition

$$\zeta(s) = \sum_{n=1}^{\infty} \frac{1}{n^s} \quad (s > 1)$$

and Euler's theorem connecting the zeta function to the primes is

$$\zeta(s) = \prod_{p \text{ prime}} \left(1 - \frac{1}{p^s}\right)^{-1}$$

(Since there are infinitely many primes, the product here is an infinite one.)

SEVEN

Geometry Without Pictures

The Hodge Conjecture

On the principle that an author should delay as long as possible introducing anything that is likely to make his reader give up in despair, when I numbered the seven Millennium Problems for the purpose of writing this book, I put the Hodge Conjecture last. In fact, my entire ordering of the problems is not how the Clay Institute listed them. Their list was based on the length of the titles, starting with the shortest (the P=NP Problem) and ending with the longest (The Birch and Swinnerton-Dyer Conjecture), so that the problem list formed an attractive Christmas tree shape on the original announcement poster:

P versus NP
The Hodge Conjecture
The Poincaré Conjecture
The Riemann Hypothesis
Yang–Mills Existence and Mass Gap
Navier–Stokes Existence and Smoothness
The Birch and Swinnerton-Dyer Conjecture

I have used a different ordering for two reasons. First, I wanted to ensure that later chapters could make use of material introduced earlier. My second reason was to ensure that the problems that are more difficult to understand come later. (It is virtually impossible to say which problems will be the more difficult to solve. All seven are on the list because they are generally acknowledged to be among the hardest unsolved problems around.)

Thus, if you're feeling pleased with yourself for having got this far (even if you had to bail out halfway through Chapter 6), and within a page or so from now you have a sudden sinking feeling that you just aren't getting it, please don't despair. In fact—and this is not something I say often—if you find the going too hard, then the wise strategy might be to give up. The Hodge Conjecture, which was formulated in 1950 by the British mathematician Sir William Vallance Douglas Hodge, is easily the least accessible of all the Millennium Problems. As an author trying to find a way to explain the Millennium Problems to a nonmathematical reader, this one caused me by far the greatest difficulty. It is a highly technical question, buried deep in a forest of highly abstract advanced mathematics known to few professional mathematicians. It deals with objects that are so far removed from the intuitions of even the experts that not only is there no "smart money" on whether the conjecture will turn out to be true or false, there isn't even a consensus as to what it really says.

You don't believe me when I say it's much worse than the other six problems? Here is the Hodge conjecture:

> Every harmonic differential form (of a certain type) on a non-singular projective algebraic variety is a rational combination of cohomology classes of algebraic cycles.

Anyone who understood just one of the technical terms in that sentence may go to the head of the class.

Okay, that was a bit unfair. I could have created a similar feeling of bewilderment by stating some of the other Millennium

Problems in a technical way right at the start. The difficulty is, with this problem, it's nearly impossible to explain what all of those technical terms signify.

Admittedly, things were getting pretty hairy with the Birch and Swinnerton-Dyer Conjecture in the previous chapter. But at least in that case I was able to relate the problem to a simple geometry puzzle. Thus, even if (like most readers, I suspect) you found the going tough —you might even have given up—at least you could start to approach the problem. You could, for instance, have told yourself that the conjecture was a more general version of the triangle area problem that arises when you first reformulate the geometry question in terms of algebraic equations and then look at all equations of the same general form. Looking at it this way definitely leaves you far short of the understanding the expert has; but the general sense you get is correct.

With the Hodge Conjecture, there is no similar path even to the problem's front door. Whereas it's just a single step from the understandable problem (the algebraic version of the area question) to the Birch and Swinnerton-Dyer Conjecture (albeit a step that involves a lot of heavy-duty mathematics), the same is not true for the Hodge Conjecture. It takes several steps to arrive at the conjecture starting from mathematical concepts everyone meets at high school. And they are steps that most professional mathematicians find daunting.

The Hodge Conjecture illustrates perhaps most clearly of all the Millennium Problems the point I raised in Chapter 0, that the nature of modern mathematics makes much of it all but impossible for the layperson to appreciate. For a century now, mathematicians have built new abstractions on top of old ones, every new step taking them further from the world of everyday experience on which, ultimately, we must base all our understanding. As I have observed before, it is not so much that the mathematician does new things; rather, the objects considered become more abstract—abstractions from abstractions, and abstractions from abstractions from abstractions. In the case of the Hodge Conjecture, the operations of calculus play a major role (differentiation, integration, etc.). But the calculus is not done on the real numbers, as many high school students learn it, or even on the complex numbers. It's calculus done in a much more general, more abstract setting.

To the layperson, the very inaccessibility of the problem is perhaps its most interesting feature. A hundred years ago, any problem in mathematics could be explained to an interested layperson. Today, some problems cannot be explained even to most professional mathematicians.

The human brain has to work hard to achieve a new level of abstraction. Only when one new level has been mastered is it possible to abstract from that level to yet another level. This is part of the reason why it takes so many years for a young mathematician to reach the frontiers of certain branches of the subject. (There are a few areas where this is not the case, but they seem to be diminishing, as mathematicians continually find ways to use techniques from the more abstract areas to advance the frontiers of the seemingly more familiar, such as the application of the calculus of complex functions to prove theorems about prime numbers that we encountered in Chapters 1 and 6.)

Having said all this, I shall nevertheless try to explain what the Hodge Conjecture says. It is inevitable that some of what I say will annoy the experts. But then, they don't need this book, do they?

The Hard Stuff, Made as Easy as I Can

In the seventeenth century, the French philosopher René Descartes showed how to reduce geometry to algebra. Instead of talking about a straight line in the plane, you could refer to the set of all points (x, y) that satisfy an equation such as

$$y = 3x + 7$$

or

$$y = \frac{3}{4}x - \frac{1}{2}$$

or whatever. (The first equation gives the straight line that passes through the point $(0, 7)$ and has slope 3; the second equation determines the line through the point $(0, -\frac{1}{2})$ having slope $\frac{3}{4}$.) Again, instead of talking about a circle you could refer to the set of all points (x, y) that satisfy an equation like

$$x^2 + y^2 = 5$$

or

$$(x - 3)^2 + (y - 5)^2 = 81$$

(The first equation gives the circle with radius $\sqrt{5}$ and center the origin; the second defines the circle with radius 9 and center the point $(3, 5)$.) Using Descartes's approach, the geometric and logical arguments favored by the ancient Greeks for solving geometric problems could be replaced by doing algebra—solving equations. Geometry done using algebra is generally referred to as algebraic geometry, or sometimes Cartesian geometry, in honor of Descartes.

During the nineteenth century, mathematicians took Descartes's approach a step further. Instead of using algebra simply as a tool to help them reason about geometric objects—by writing down equations that determined those objects—they started with collections of algebraic equations and *defined* "geometric" objects to be the solutions to those equations. Thus, instead of saying that the equation

$$x^2 + y^2 = 4$$

provides an algebraic description of the circle with radius 2 and center the origin, they would simply study the object—whatever it is—that arose from that equation. In the case where we start out with a familiar geometric object, this gives nothing new of course, apart from a different perspective on what comes first. But most equations do not correspond to familiar geometric objects. So it made no sense to call them "geometric objects." The name mathematicians gave to objects that arise from algebraic equations in this way was "algebraic variety." Actually, that's not exactly right. In defining an algebraic variety, mathematicians don't restrict themselves to a single algebraic equation; instead, you can start with any finite collection of equations. The variety then consists of all points that solve all the equations in the system. This makes the class of algebraic varieties richer than if you were only allowed to start with a single equation. (In the case of a system of two equations, each of which defines a familiar geometric figure, the variety determined by the system will be the intersection of those two figures—the parts that lie on both figures.)

Thus, an algebraic variety is a generalization of a geometric object. Any geometric object is an algebraic variety, but there are many algebraic varieties that cannot be visualized. Just because a particular algebraic variety cannot be visualized, however, it doesn't follow that you cannot do (algebraic) geometry on it. You can. It is geometry without pictures.

Now we can take a look at one of the technical terms in the Hodge Conjecture: A nonsingular projective algebraic variety is, roughly speaking, a smooth multidimensional "surface" that results from the solution of an algebraic equation. (Much as a sphere is a smooth two-dimensional surface that results from solving the algebraic equation $x^2 + y^2 + z^2 = a^2$, for some number a.)

The conjecture makes a claim about "harmonic differential forms" on those "surfaces." Loosely speaking, a harmonic differential form is a solution to a certain very important partial differential equation called Laplace's equation, that arises both in physics and in the study of functions of complex numbers. (I'll state this equation later.)

Now, when students first learn calculus at university, they usually do it on the two-dimensional plane, the familiar plane surface of Euclid's geometry and elementary trigonometry. But with a little effort, calculus can be developed on other surfaces, for example on the surface of a sphere. With a lot of effort, you can (more precisely, the experts can) develop calculus on much more general kinds of varieties. The Hodge conjecture deals with calculus developed on a nonsingular projective algebraic variety. It makes a claim about certain kinds of abstract objects, let us call them H-objects, that arise when you start with a certain kind of variety and do some calculus on it.

Now, when you use calculus (as opposed to algebra) to define an object, the resulting object need not be in any sense "geometric." The Hodge Conjecture says that the H-objects are exceptions to that last remark—or at least, almost so. Though they may not themselves be geometric objects, they can be built up from geometric objects in a fairly straightforward (and calculus-free) way. In the terminology of the conjecture, an H-object is a rational combination of cohomology classes of algebraic cycles. That is to say, any H-object can be built up from geometric objects in a purely algebraic way.

Thus, you can think of the Hodge Conjecture as saying, "Look, by using calculus on varieties, we've created a class of objects (the H-objects) that not only defy any hope of our being able to visualize them, we can't even describe them algebraically. However, these objects can be built up in an algebraic fashion from objects that can be described algebraically. So at least we still have a lifeline connecting us to firmer ground—a connection that we (i.e., the experts) might be able to use to carry the study of these objects further."

What the Hodge Conjecture does is provide the expert with some powerful mathematical structure that can be used to analyze H-objects. This is very typical of a lot of modern mathematics, where mathematicians are constantly looking for new structures on objects or links from one area or mathematics to another, so that they can adapt methods from one area for use in another. (This is, after all, exactly what Descartes did when he showed how to use the methods of algebra to study geometric objects.)

Now, those last two paragraphs consist of the kind of statements that drive the experts wild. In order to provide at least a general sense of what is going on, I am trying to put into familiar language something that is so far removed from the everyday world that any such attempt is, in a strict sense, doomed to fail. But let's press on anyway. Here is another way of getting at the problem that has the advantage of involving notions that won't seem too strange to anyone who has taken a college calculus course, but which has the disadvantage that, while technically correct, it "misses the point" of the Hodge Conjecture.

We can formulate the Hodge Conjecture by starting with integrals over generalized paths on algebraic varieties. Deforming the paths leaves the values of such integrals unchanged, so you can think of the integrals as being defined on classes of paths.

The Hodge conjecture proposes that if certain of these integrals are zero, then there's a path in that class that can be described by polynomial equations.

As I mentioned, this formulation of the Hodge Conjecture is technically accurate but misses its main spirit. Before trying to convey some sense of the way the experts view the conjecture, let me make a few more remarks about its status in mathematics.

First, it has significant implications. A proof of the Hodge Conjecture would establish a fundamental link among the three disciplines of algebraic geometry, analysis, and topology.

Second, one case of the conjecture is known, but that case was established by the American mathematician Solomon Lefschetz in 1925, long before Hodge formulated the general conjecture. It's not much of an exaggeration to say that there has been essentially no progress on the problem since then.

Thus, to date, the Hodge Conjecture remains just that: a conjecture. Some would say it could be more accurately called a wild guess. But that has not prevented many mathematicians from trying to prove it or from investigating its consequences. In 1991, the American Mathematical Society published a book cataloguing some of the research that had been done on the Hodge Conjecture.[1] A second edition was brought out in 1999, updating what was known. That second edition is 368 densely packed pages long. It includes a new section that lists 71 papers

---

1. James D. Lewis, *A Survey of the Hodge Conjecture*, CRM Monograph Series, Volume 10, Providence, RI: American Mathematical Society.

published between 1950 and 1996 on just one aspect of the conjecture, the Hodge Conjecture for so-called abelian varieties. In the preface, the book's author admits that even with this addendum, the survey is still not complete, and refers readers to other sources.

Incidentally, here is how that American Mathematical Society volume states what it describes as the "popular version" of the Hodge Conjecture in the opening paragraph of the preface:

Let X be a projective algebraic manifold and p a positive integer. Also, let $H^{2p}(X, \mathbb{Q})_{\text{alg}} \subset H^{2p}(X, \mathbb{Q})$ be the subspace of algebraic cocycles, i.e., the $\mathbb{Q}$-vector space generated by the fundamental classes of algebraic subvarieties of codimension p in X. The Hodge conjecture asserts that one can "compute" the subspace $H^{2p}(X, \mathbb{Q})_{\text{alg}}$ using Hodge theory, specifically $H^{2p}(X, \mathbb{Q})_{\text{alg}} = H^{p,p}(X) \cap H^{2p}(X, \mathbb{Q})$.

So now you know.

If you are still with me, you're ready to dig a little deeper into the conjecture and see how it came about.

Who Was William Hodge?

For a mathematician who played such a prominent role in the profession, surprisingly little is known about William Hodge. His life remains as inaccessible as his conjecture.

He was born in 1903 in Edinburgh, Scotland. A brilliant student, after studying at Edinburgh and then at Cambridge, in 1936, at the tender age of 33, he was appointed to a chair at Cambridge, which he held until his retirement in 1970.

He was one of the leading figures in developing the relationship between geometry, analysis, and topology. Mathematicians today remember him mostly for (besides his conjecture) his theory of harmonic integrals.

He was elected to the Royal Society of London in 1938, and awarded the Society's prestigious Royal Medal in 1957 (in "recognition of his distinguished work on algebraic geometry"), and knighted by the Queen in 1959. He served as president of the London Mathematical Society from 1947 to 1949 and won their Berwick Prize in 1952. In 1974, the Royal Society honored him once again, this time with the Copley Medal, citing "his pioneering work in algebraic geometry, notably in his theory of harmonic integrals." As an active promoter of mathematics both in Britain and abroad, Hodge was one of the originators of the British Mathematical Colloquium (an annual conference that visits different British universities) and played a major role in setting up the International Mathematical Union in 1952. He died in 1975, aged 72.

Having devoted most of his career to developing the deep and rich theories of algebraic geometry in which his conjecture arises—one of which is now called simply "Hodge theory"— Hodge announced the conjecture itself in his address at the 1950 International Congress of Mathematicians held at Cambridge, England.[2] (More precisely, in his lecture Hodge stated it as an open problem. However, other writings he published on the subject indicate that he believed it was likely to be true.)

Now that you know as much about Hodge as anyone else, let's take a look at the various developments in mathematics that led up to his formulation of what is now a $1 million Millennium Prize Problem.

When Complex Numbers Met the Mathematics of Fluids

The story begins in Italy during the Renaissance, where mathematicians began to talk about doing the unthinkable: introducing into algebra a number whose square was -1. This is the number mathematicians now denote by the letter i, and which, as we saw in Chapter 1, forms the basis for the complex numbers.

We also saw in Chapter 1 that although the human mind finds it difficult at first to accept the idea of a number whose

---

2. See W. V. D. Hodge. *The topological invariants of algebraic varieties*, Proc. International Congress of Mathematicians (Cambridge, 1950), Amer. Math. Soc., Providence, RI, pp. 182–192.

square is negative, the complex numbers nevertheless have an arithmetic that works just like the ordinary arithmetic of real numbers. You can add, subtract, multiply, and divide two complex numbers, and you can solve polynomial equations involving complex numbers. We saw too that the way to overcome the counterintuitive nature of complex numbers is to realize that they can be pictured as the points on the ordinary two-dimensional plane: The complex number $x + iy$ is (or, if you prefer, is represented by) the point in the plane with coordinates (x, y).

Now, the real numbers can be paired off in a natural way, by associating each real number r with its negative $-r$. If we picture the real numbers as the points on a line (the "real line"), then the pairing has an elegant representation where each number is paired with the number an equal distance from the origin on the opposite side of the origin. (See Figure 7.1.) This particular pairing plays a major role in the arithmetic of real numbers. (For example, students at school learn that to solve an equation you often add to both sides the negative of one of the terms in the equation.)

Complex numbers can be pictured as points on a plane (the complex plane). For these numbers, the analogous pairing that takes $x + iy$ to $-x - iy$ is a reflection in the origin. But there is another pairing of complex numbers that turns out to play a major role in complex arithmetic. This second pairing associates with each complex number $x + iy$ its *complex conjugate $x - iy$*. Figure 7.2 gives a geometric illustration of this pairing. Just as the real-number pairing is reflection in the origin of the real line, so too the complex conjugation pairing is reflection in the real axis (i.e., the x-axis) of the complex plane.

Despite some initial resistance to using complex numbers, by the nineteenth century their basic theory had been well worked

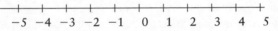

Figure 7.1. The real numbers form pairs, which differ only in sign, that are symmetric about the origin.

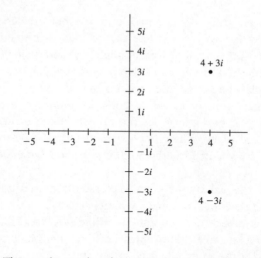

Figure 7.2. The complex numbers form pairs, called complex conjugates, that are symmetric about the real axis. The conjugate pair $4 + 3i$, $4 - 3i$ is shown.

out, and they were generally regarded as the standard number system of mainstream mathematics. Moreover, mathematicians started to develop a deep and beautiful generalization of calculus to complex functions, giving the subject known today as complex analysis.

Two of the leading figures in early complex analysis were Bernhard Riemann, whom we have met several times already, and Augustin-Louis Cauchy. They made a dramatic and unexpected discovery connecting complex functions with physics. They began by saying to themselves, "If $f(z)$ is a complex-valued function of a complex variable z, then we can write the value $f(z)$ of this function in the form $f(z) = u(z) + iv(z)$, where $u(z)$ and $v(z)$ are real numbers. This gives us two new functions u and v, both *real-valued* functions of a complex variable z." (Mathematicians today call the functions u and v the real and imaginary parts of the function f.)

The two mathematicians discovered that if the complex function f has a well-defined (calculus) derivative—in modern terminology, if the function f is *analytic*—then its real and imaginary parts u and v have to satisfy the two partial differential equations

$$\frac{\partial u}{\partial x} = \frac{\partial v}{\partial y} , \quad \frac{\partial u}{\partial y} = -\frac{\partial v}{\partial x}$$

These equations are (and were then) familiar to physicists. They know them as Laplace's equations, and they play a major role in gravitational theory, electromagnetic theory, and fluid mechanics. (Moreover, closely related equations occur in the theories of heat flow, acoustics, and the propagation of waves.) A solution of the Laplace equations is called a harmonic function. The discovery of this close relationship between the calculus of complex functions and the Laplace equations led to significant progress in mathematical physics, by providing a way to solve Laplace's equations in a variety of contexts.

One important development in the theory of complex functions was Riemann's invention of what are nowadays called Riemann surfaces. There are some functions that work just fine for real numbers, but when the argument or the value is allowed to be a complex number, the result isn't a proper function at all, because a single argument can lead to more than one value. The square root function and the logarithm are two examples. For real numbers, any positive real number has two square roots, but because one of those is positive and the other negative, the problem can be eliminated by simply choosing the positive root; indeed, the standard symbol $\sqrt{a}$ is always understood to refer to the positive square root of a. But when the root is a complex number, there is no natural and useful way to choose between the two roots. Riemann proposed that the best way to handle these "many-valued functions" (which are not really functions at all) is to think of them as single-valued functions (i.e., genuine functions) defined on a multilayered surface. Riemann surfaces have a more complicated (and more interesting) topology than the complex plane. One way to think of them is as a spiral-staircase configuration of complex planes, stacked one on top of the other. Each complete counterclockwise rotation around the origin in one sheet of the surface takes you onto the next sheet up. (See Figure 7.3.)

The Hodge Conjecture: Not for the Faint-Hearted

In the early twentieth century, mathematicians generalized the idea of a Riemann surface to the highly abstract concept of a complex manifold, a multidimensional analogue of a Riemann surface with a complicated topology. Such a manifold is equipped with a structure that ensures that the concept of a complex analytic function makes sense. In particular, it is possible to define what are called differential forms, generalizations to many dimensions of the differential df of a function f in ordinary (real number) calculus.

Some of these differential forms fall into distinct classes having certain key features in common, not unlike the way humans fall into distinct national groupings sharing a common language, history, and culture. Because of the way those classes arise, they

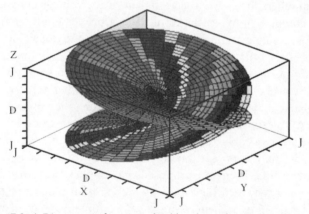

Figure 7.3. A Riemann surface, a surfacelike object that covers the complex plane with two or more (in general, infinitely many) "sheets." The figure shown has just two sheets that interconnect as they wrap around the origin, but in general the sheets can have very complicated structures and interconnections. Riemann surfaces are one way of representing multiple-valued functions.

are called cohomology classes. These cohomology classes are what the Hodge conjecture talks about.

It requires a chain of highly technical mathematics to get to the notion of a cohomology class, and that would take far too long to cover here. But for the record, here is a very brief summary.

- First, we need to know that there is a particular operation on differential forms, called the exterior derivative. The exterior differential is itself a kind of differential.
- A differential form is said to be closed if it is the exterior derivative of some other differential form.
- A differential form is said to be exact if its own exterior derivative is zero.
- Two closed differential forms are called cohomologous if their difference is exact.

Thus, the members of the cohomology classes are closed differential forms. Exactness is the "similarity" property that members of the same cohomology class share. Notice that the definition of the cohomology classes depends heavily on notions from calculus.

The cohomology classes define useful topological invariants that capture important aspects of the underlying complex manifold.

With the idea of a cohomology class (of closed differential forms) under our belts, we can go back to algebraic geometry and the notion of an algebraic variety.

A complex algebraic variety is a multidimensional "surface" defined by the complex solutions of a system of algebraic equations.

Mathematicians say that a complex algebraic variety is projective if the solutions of the defining equations depend only on the ratios of the numbers involved.

They say that the variety is nonsingular if the surface is smooth.

Thus a nonsingular projective complex algebraic variety is a special kind of complex manifold.

Hodge realized that he could apply methods from analysis to these algebraic manifolds. Specifically, he realized that rational cohomology classes of differential forms that arise from a nonsingular projective complex algebraic variety can be viewed as solutions of the Laplace equations.

Hodge's observation makes it possible to write such a class as a sum of special components, what are called the harmonic (p, q)-forms. These are solutions of the Laplace equations that can be specified by p complex variables and q conjugate complex variables. Moreover, every algebraic cohomology class (of dimension p) gives rise to a (p, p)-form.

In his address to the International Congress of Mathematicians in 1950, Hodge raised the possibility that for nonsingular projective complex algebraic varieties, that last property completely characterizes the algebraic cohomology classes. That is, every harmonic (p, p)-form is a rational combination of closed algebraic forms (loosely, could be built up in an algebraic—i.e., calculus-free—way).

Thus was the Hodge Conjecture born.

But is the conjecture true? Nobody knows. At the moment there is no strong evidence to suggest that Hodge's intuition was correct. On the other hand, even when the mathematics is at its most abstract and esoteric—and the Hodge conjecture certainly fits that description—a trained human mind that has thought long and hard about a particular problem frequently develops intuitions that prove to be correct. Hodge knew the material I have just sketched intimately. Indeed, he developed much of it. Personally, I would not be the least surprised to learn that he was correct. And that would say more about the mysteries of the human mind than about harmonic (p, p)-forms.

Further Reading

Anyone who is contemplating making a serious attempt to solve one of the Millennium Problems should be aware that the descriptions of the problems given in this book are loose ones, designed to give an overall sense of the issues. For precise descriptions of the seven problems, together with the official rules for the competition, consult the Clay Mathematics Institute website at www.claymath.org. (The website also features a twenty minute streaming video on the Millennium Problems, presented by Keith Devlin and produced by the Clay Institute.) In addition, the Clay Institute and the American Mathematical Society are planning to publish jointly a book in which each of the seven problems is described in detail by a world expert on the problem. (If you cannot follow the descriptions in that book, you are unlikely to be able to solve any of the problems.)

If, on the other hand, you are content just to know, more or less, what are the seven unsolved problems that present-day mathematicians think are the hardest, and your desire now is simply to delve further into the world of mathematics, then there are a number of excellent books you can consult, written for a general audience. Most of them, however, are either written at a much more superficial level than the present one, or else are focused on specific issues in mathematics or on historical issues. There are relatively few general audience books that give a good overview of contemporary mathematics and are written at roughly the same level as the present one. The ones I am familiar with are:

Arnold, V., M. Atiyah, P. Lax, and B. Mazur, eds. *Mathematics: Frontiers and Perspectives*. the American Mathematical Society, 1999.

Casti, John. *Five Golden Rules: Great Theories of 20th Century Mathematics—and Why They Matter*. John Wiley & Sons, 1996.

Devlin, Keith. *Mathematics: The New Golden Age*. Columbia University Press, 1999.

Devlin, Keith. *The Language of Mathematics: Making the Invisible Visible*. W. H. Freeman, 1998.

Yandell, Benjamin. *The Honors Class: Hilbert's Problems and Their Solvers*. A. K. Peters, 2001.

Index